The Calm Birth Method

The
Calm
Birth
Method

Your Complete Guide to a Positive
Hypnobirthing Experience

SUZY ASHWORTH

HAY HOUSE

Carlsbad, California • New York City • London
Sydney •Johannesburg • Vancouver • New Delhi

First published and distributed in the United Kingdom by:
Hay House UK Ltd, Astley House, 33 Notting Hill Gate, London W11 3JQ
Tel: +44 (0)20 3675 2450; Fax: +44 (0)20 3675 2451; www.hayhouse.co.uk

Published and distributed in the United States of America by:
Hay House Inc., PO Box 5100, Carlsbad, CA 92018-5100
Tel: (1) 760 431 7695 or (800) 654 5126; Fax: (1) 760 431 6948
or (800) 650 5115; www.hayhouse.com

Published and distributed in Australia by:
Hay House Australia Ltd, 18/36 Ralph St, Alexandria NSW 2015
Tel: (61) 2 9669 4299; Fax: (61) 2 9669 4144; www.hayhouse.com.au

Published and distributed in the Republic of South Africa by:
Hay House SA (Pty) Ltd, PO Box 990, Witkoppen 2068
info@hayhouse.co.za; www.hayhouse.co.za

Published and distributed in India by:
Hay House Publishers India, Muskaan Complex, Plot No.3, B-2,
Vasant Kunj, New Delhi 110 070
Tel: (91) 11 4176 1620; Fax: (91) 11 4176 1630; www.hayhouse.co.in

Distributed in Canada by:
Raincoast Books, 2440 Viking Way, Richmond, B.C. V6V 1N2
Tel: (1) 604 448 7100; Fax: (1) 604 270 7161; www.raincoast.com

A catalogue record for this book is available from the British Library.

ISBN: 978-1-78180-846-7

Interior images: Suzy Ashworth

Printed and bound by CPI Group (UK) Ltd, Croydon, CR0 4YY

To Jerome, Caesar, Coco and Aluna –
my beautiful family,
I love you so much. xxx

Contents

Contents

Acknowledgements

The biggest thank you has to go to my long-suffering husband, Jerome Ashworth. I'm so grateful for all his support and belief in me. During the many long nights and weekends, he urged me to keep on going and for that I will be forever thankful! My beautiful children, Caesar, Coco and Aluna, who won't know for a long time what they have inspired, but without whom there would be no Calm Birth School.

To all The Calm Birth School instructors who help to spread The Method and teachings to women far and wide. The godmother of hypnobirthing as we recognize it in the West, Mary Mongan. The founders of The Wise Hippo, Dany Griffiths and Tamara Ciafinni, who are a constant inspiration with their work to empower women and their families to create calm and positive birth experiences.

To Mark Harris, father, midwife and author of *Men, Love & Birth* for the support he offered The Calm Birth School when it was barely a thought.

To Katy Appleton of Apple Yoga, Katie Whitehouse from Vital Touch and the Neighbourhood Midwives for their ongoing support.

To all the couples who have been through The Calm Birth School and have kindly shared their journey both within and off the pages of this book; they are a constant inspiration and motivate me to do more.

To Elloa Atkinson, for believing in the message of this book and her role in helping me to get to this point.

To all the childbirth educators out there who continue to make mothers and babies their priority, I thank you, too.

Introduction:
No Vagina Whispering

So you've finally decided to bite the bullet, or maybe some kind soul decided to for you, and you want to know more about The Calm Birth Method (TCBM): the tools and techniques used in the world's first video-based hypnobirthing programme, The Calm Birth School. Since its launch in December 2014, The Calm Birth Method family has helped over 5,000 women across six different continents to create calm and positive birth experiences, and now has a team of amazing instructors working with women face to face across the whole of the UK, so you're in the best company. But enough about us, let's cut to the chase and acknowledge the most important person in the room: that's you. Yes, you, who right now is growing another human being, even as you read these words. And that is by far the biggest deal in the world. You are bloody amazing.

Perhaps the whole 'growing a baby' thing is something that you've already got to grips with. If not, go and Google it. There's a million-and-one blogs and forums telling you exactly what size nut your baby is today. This book isn't about any of that, but rather the most amazing feat a woman's body ever goes through – giving birth. In the following chapters you'll get the lowdown on all the different tips, tricks and techniques that have helped thousands of women to have calm, empowering, positive and fear-free birth experiences.

Although this is primarily a book about giving birth, you'll also learn how to unleash your personal power from within, so that when the moment finally arrives for you to welcome your baby into the world, you'll feel calm, positive and ready for whatever twists and turns may be presented to you. The one thing we know for sure about birth is that it's a journey, and one where you can't know ahead of time exactly what's going to happen. You can, however, arm yourself with a mindset and a toolkit that can make giving birth one of the most incredible and enjoyable experiences of your life.

By the time you have finished reading this book, you'll understand exactly what you need to do to give yourself permission to go wherever you need to go, controlling what you can control and letting go of what you can't, as you eliminate the fear of birth. It's this inner confidence that will enable you to bring your baby into the world positively, happily and powerfully.

Am I serious about birth being happy, positive and enjoyable? Absolutely. Am I a lentil-eating, hummus-knitting, crochet-knickers-wearing kind of woman? Perhaps you'll be surprised to hear, I'm not. Am I going to make you walk from room to room, wafting a joss stick while chanting in Sanskrit? Nope. Will I be asking you to pay homage to the lotus flower or whisper to your vagina? No, no and no: there will be no vagina whispering! (Unless, of course, you want there to be.)

While I've secretly loved all things a bit hippy for a while now, this book was written from my shoot from the hip, let's get sh*t done, pragmatic left-hand brain. I write these words to you as an everyday woman who has been there and got the calm birth T-shirt, not once or twice but three times. And I'm here to show you that no matter who you are, or what your perspective on birth, giving birth can be an experience you'll never forget, for all the right reasons.

While you might be feeling confident in your body's ability to do this job (and why the hell not, it is quite literally what your body is designed to do), it is also highly possible that you're peeing your pants at the thought of everything to do with the bit before you hold your baby in your arms for the first time.

If this is your first child and you're anything like me, or many of my clients, you might be in a bit of a head spin about having a baby growing inside you. That's OK. It's normal to get a bit freaked out as you comprehend that the collection of cells formulating within you right now – which started out smaller than a grain of rice – will one day become an actual human being. It definitely fried my brain – even third time around!

Or perhaps the mere thought of squeezing a nine-pounder out of your lady parts brings tears to your eyes... so you aren't thinking about it. Maybe you're so busy being superwoman that the fact you're about to become a mother and give birth has barely registered on your inner Richter scale. You're just too goddamn busy right now to give it any headspace. Or, maybe you're the control freak who wants to know all the details. For you, the thought of your body doing things you can barely bring yourself to imagine makes you want to reach for a large gin, so you're arming yourself with as much information as you can get. As I've said before, we at The Calm Birth School have been there and done this, and we work with women like you every day. I want you to know that it doesn't matter who you are or whether or not you're experienced or knowledgeable about birth: you really have got this.

By reading this book, you're acknowledging that at least a small part of you knows birth doesn't have to be like the horror stories that are shown on TV, discussed with girlfriends or overheard around the water cooler. You don't have to be that disempowered woman who feels scared, vulnerable and as though her birth is happening to her. You don't have to go into labour unaware of all the things you don't know just because you haven't done this before.

The Calm Birth School's mission is to arm you with knowledge and empower you with confidence, so that you know from the core of your being that you were designed to do this. I'm here to remind you that your innate ability to give birth – which has been genetically programmed into all of womankind over hundreds of thousands of years – is right there at your fingertips.

Will this guarantee you an easy, straightforward birth during which your baby pops out reciting the alphabet backwards? (Hypno-babies are very advanced, you know.) Sadly, no. I wish I could guarantee such linguistic feats alongside a 'perfect' birth, but that isn't the case.

So what is the point of hypnobirthing then?

I'd like to invite you to think of your birth as the most important meeting of your life to date. What would you do before you knocked on the office door, ready to strut your stuff in front of your boss? You would prepare everything down to the nth degree. You would have every base covered so that whatever curveballs you may be faced with, you'd know you could handle it all. You'd feel calm, confident and ready to go.

You already know that as first meetings go, the date you have with your baby is going to be infinitely more impactful, meaningful and enjoyable than any meeting with a boss. My goal is for you to feel even more confident, prepared and knowledgeable than you would be if you were prepared to come face to face with the people who pay your wages.

Why is this so important? Well, contrary to popular opinion, I will unreservedly declare that giving birth should be one of the most empowering, life-affirming and joyful events you'll ever experience. When you learn to work with your body, as opposed to against it, you'll know how to embrace the sometimes weird, sometimes wonderful power and intensity that moves up, down, around and through your body. And when you allow yourself to go with that flow, knowing with every fibre of your being that you were born to do this, then yes, it really can feel amazing.

✍ Rachel's birth story ✍

Yesterday, our darling baby girl, Wynter, arrived and it was truly the most amazing calm birth experience. I'm a first timer, so I didn't expect things to progress as quickly as they did. And I really didn't expect to end up having my dream birth.

My waters released at 1:45 a.m. and a few hours later I was in the midwife-led birthing unit at the Princess Alexandra Hospital and 3cm dilated. Breathing and TENS took me to 6–7cm, and then I got into the birthing pool for the final part of the journey. Wynter arrived safely at 10:45 a.m.

My husband was an excellent birth partner, and helped me to stay strong enough to labour without any pain relief apart from gas and air.[1] He said he'd never seen me so calm and focused. And I still feel calm now, and Wynter is so relaxed that we have to wake her up for feeds. In other words, all is well and I'd like to thank you for helping me to have the birth I wanted.

The calm birth experience is for you, of course, but it's also for your baby. When you release the resistance that can impede labour and birth, your baby's experience of entering the world becomes infinitely less jarring – a calm and positive end to your pregnancy and start to the newest chapter of your family's life. You can then focus on the most important job of all: getting to know this new little person who has just arrived. A positive, calm birth will also promote attachment and bonding with your baby, support you in being completely present at every precious moment of your newborn's life, and give you the emotional space that all new parents need to navigate the transition into parenthood.

The flipside of the above scenario is the negative or stressful birth experience, which unfortunately I hear about all too often from second-time parents. Research shows that negative experiences during childbirth can lead to difficulties with bonding[2] and establishing breastfeeding.[2-3] I have seen hurt, anger and trauma from women who, when they look back at their birth experience, feel that they were ignored or disempowered. This is often coupled with feelings of isolation and guilt at not being able to express the disappointment of their experience, for fear of appearing ungrateful for the healthy baby they're holding in their arms. I hear stories of stress and angst from both mums and dads as they try to make their way through the negative roller-coaster of emotions that a stressful birth experience can trigger.

The message of this book is simple: **birth doesn't have to be like that.**

In this book, you'll learn:

- Simple but powerful tools and techniques to help you feel calm, confident and at ease during labour.

- How the mind–body connection can affect your ability to give birth optimally, and what you can do during your pregnancy to get it working for you, rather than against you.

- How to embrace your pregnant self and why it is important for your birth.

- How you can get your care providers working with you for your specific needs, so that you feel like you're the one running the show during your pregnancy and birth, no matter where and how you give birth.

- Some great ways to get your birth partner involved in your pregnancy, so they feel equipped to support you and be your advocate on the day.

I can't emphasize how important your birth partner will be during this process, so if you know it's going to be an uphill challenge getting them to read this book with you, I highly recommend The Calm Birth School video course. This has all the content split up into 10- to 20-minute videos that you can watch together in the comfort of your home. It feels even more like you're getting a private class from me and this, along with the personalized support in the students' Facebook group, means you really can't go wrong.

Check out all the details for the video programme here: www.thecalmbirthschool.com/course

Whether you read this book alongside studying the course or not, you'll finish this book feeling uber-prepared for the most important day of your life.

Ready to get started?

Let's do this!

Suzy xo

How to Use This Book

Maybe you have a huge pile of pregnancy and birth books on your bedside table, or perhaps this is the only book you intend to read before your baby arrives. (There are a million other things you could be doing, right? After all, reading lots of pregnancy and birth books makes everything a bit too real, doesn't it?) Either way, this is the most important book you'll read about creating a positive birth experience.

Within the following chapters and the wonderful Calm Birth School community on Facebook, you'll not only find the tools you need for an amazing birth, but also a philosophy to take into your life. You'll gain a better understanding of what you can control and what you can't, develop an enhanced appreciation of the difference between flow and resistance, and understand what you need to do to adjust to your environment to suit you and your baby. Starting today, and while reading this book, you'll gain a newfound inner confidence and experience your unique ability to tune in to the changes within your body as it grows with your baby.

This will make it even clearer to understand exactly what you and your family will require to create the best possible foundation for the most positive birth experience that you deserve and desire.

To get the most out of this book, it is best to read each section, practise the exercises and do homework as you go along. While it might be tempting to skim-read, solely looking for the techniques for labour, The Calm Birth Method (TCBM) takes a holistic approach to learning, so if you only look for the bits about 'how to breathe', for example, you won't do yourself or your birth justice. I wish it could be as easy as saying 'read and breathe and you'll be fine', but it doesn't work like that. I know that when you practise these techniques they will work, and not just because I have used them during my own births, but also thousands of other women have successfully used this exact same process, too.

The line in the sand is clear when it comes to understanding those women who benefit most from these tools. The secret is, it's not really a secret. The success stories come from the women who put time into completing the homework and tuning in to the belief that they can create a positive birth experience, regardless of whether their baby decides to follow their birth preferences to the letter or has other ideas.

The clients who asked questions, invested in The Calm Birth School course, bought books, did their research and, in some cases, got extra support when they needed it, focused on and strengthened their ability to create a positive birth experience.

Please note that at no point will I say that positive equates to natural. Your birth doesn't have to be quick, natural or textbook for you to have an amazing experience. As you'll see from the array of birth stories in this book – which include C-sections, home births, hospital births, long and very quick births, and everything in-between – each woman's experience is different.

And it all starts with how you think and speak about giving birth to your baby.

The language of birth

The language you use and hear about birth is especially important because it can make a real difference to whether you feel calm or anxious about the prospect of giving birth.

The reason for this is that words have the power to evoke huge emotional responses. A kind word can make the stress of a horrible day fade in seconds, while a person who knows you well might get your blood boiling by stating something that wouldn't register with anyone else. For example, someone might innocently say to a pregnant mama-to-be, 'So are you going back to work once you've had the baby, or are you just going to be a stay-at-home mum?'

The word 'just' is often enough to trigger stress in many women, as it suggests that being a full-time mum is less worthy or important than going back to work.

Words create imagery in the mind, which in turn trigger emotions. In fact, every thought we have creates a chemical and physical response in the body, and we'll be exploring this in more detail in a later chapter. But from this point onwards, I would like you minimize your exposure to some of the more negative words or phrases about giving birth by replacing them with the more positive Calm Birth School terms, shown below:

Instead of saying...	Say...
Contraction	Wave or surge
Broken (as in waters)	Released
Pain	Discomfort, sensations or pressure
Birth canal	Birth path
Pushing	Bearing down
Complications	Special circumstances

For now, just notice how it feels to say or hear the word in the left-hand column compared to the one on the right.

Preparing for Birth

Feeling calm about the prospect of giving birth is one of the key tools of The Calm Birth School, and you'll be learning many practices in this book to help you achieve it, but one of the other key tools is being prepared.

So how are you going to prepare for your birth? If you want the kind of birth that is available for you, you'll make a commitment now to do your homework and participate in the online group.

As you go through the book, highlight all the parts that stand out as important to you and those that will be useful for your partner to read. Even though you're the one having the baby, it's really important that your birth partner gets involved too, as they will play a vital role in ensuring that everything is set up as you need it to be. Your partner also needs to have a clear understanding of TCBS techniques – what to use and when, and how to encourage and support you – without the worry of you morphing into Frankenstein's bride. We've all heard those stories about women in labour who threaten physical violence to anyone within arm's length. Your partner's preparation is vitally important and will enable them to be the best support for you, alongside providing them with their own set of tools to fall back on. Witnessing this miraculous event is no small thing and when emotions are running high, 'your' hypnobirthing tools will be equally as invaluable to your partner as they are to you.

If you're enjoying a low-risk pregnancy, hypnobirthing will help you to create a firm foundation for enjoying a natural, vaginal birth with no intervention; although of course this is not guaranteed. This is why reading the entire book is important: I want you to be at ease with the idea that you can navigate any situation you're presented with in a way that leaves you feeling positive, confident and in control of your birth, rather than it being something that happens to you.

∼ tip ∼

Listening to The Calm Birth School MP3s daily will turbocharge your learning and your confidence. In addition, you can download your free 'practice schedule' at www.thecalmbirthschool.com/bookbonuses

Once you're into the final four weeks of your pregnancy, it's a good idea to read the book again, step up your breathing and visualization practices, and you'll be good to go.

So as we prepare to dive in, remember: positive does not have to mean perfect.

∂ Laura's birth story ∂

Hazel Celia Ellerby was born on Tuesday, weighing 7lb 1oz after a very quick, natural labour of only two and a half hours.

I was 40+6 weeks and had a sweep[4] booked for 3:30 p.m. Feeling positive and calm that this might slowly kick-start the journey to meeting our baby, I was quite surprised to hear that I was 2cm dilated with a softening cervix and the baby's head was very low. The midwife hinted that I wouldn't need inducing the following week, as our baby would be out soon! Little did I know then that later that day, I would be concentrating on my breathing while Mum called the hospital to tell them that my waters had released.

The midwife advised I go to the hospital for a check-up, so Mum called Julian to tell him to meet us in the maternity unit and then a taxi. But by the time I'd made it down the stairs, I'd had two surges 30 seconds apart and felt the cramps move to my backside, and the need to push. Change of plan – get an ambulance!

The good news is that the paramedics arrived quite quickly and said I had time to get to the hospital. I was bundled into an ambulance and given gas and air, which really helped me to cope with the surges and the bumpy ride. Coincidentally, Julian met us in the lift on the way up to the ward (not that I was very aware at the time); he'd been asked to keep the elevator doors open by the paramedics as they approached, and when he looked down at the gurney he saw me!

I went into a labour room and was disappointed to see that the water-birth bath was out of action. But within moments, all my ideas of using water, aromatherapy, a beanbag and a ball went out of the window. When the midwife examined me, she found I was fully dilated with my baby's head in the perfect position for me to start bearing down during my next surge. Labour definitely didn't slow down during the journey to the hospital, then! I think I was so relieved that my mum wasn't going to be my midwife that my body just let go!

I relinquished the gas and air for the last half an hour so that I could focus on making the most of each surge to push her out. For some reason I was on my back, but when I turned onto my knees and leaned over the back of the bed, bearing down immediately felt easier. Julian was right beside me stroking my back, playing soft music on his phone and spraying lavender on the bed, so it didn't smell too much like a hospital. The lights were dimmed, too. My mum was about to leave us to it, but she'd been such a big part of the action I asked her to stay. I had a lovely midwife and an encouraging trainee, so I felt very safe.

I can't say I breathed my baby out, but the low humming while holding my breath and envisaging her coming out really helped. I reached down when I felt her head emerge, and one last push brought her out and up into my arms. I held my baby in pure wonder. We had worked together as a team; I had trusted that she knew what to do, and the 'All is calm, all is well... I'm safe' mantra played in my head throughout labour. Each surge brought her closer to me but, in the grand scale of things, there really weren't that many!

Thanks to The Calm Birth School, I had the natural, positive experience I had hoped for. And, despite it being super intense, it was a real bonus that it was so quick!

❧

Chapter 1

What Is Hypnobirthing?

When addressing the question of what hypnobirthing is, it's probably best to start with what hypnobirthing isn't.

Hypnobirthing is not hocus-pocus. As a former hypnotherapist and psychotherapist, I laugh in the face of the idea of 'mind control'... seriously, that is not hypnobirthing. The good news is that you won't find yourself barking uncontrollably every time you see a full Moon or dry-humping your birth partner in a room full of strangers for comic entertainment. Hypnobirthing has nothing to do with voodoo and everything to do with science.

So if it's not mind control or voodoo, what is it? Quite simply, hypnobirthing enables you to feel calm and connected to your baby during your pregnancy, and can provide you with invaluable tools and techniques that will allow you to stay relaxed during your birth. By the time you have completed this book – if you practise what I'm going to teach you – you'll be the equivalent of a Jedi Master at instant relaxation whatever the situation, using your breath to feel calm and at ease, as well as utilizing visualization to help increase your focus.

Alongside the tangible hypnobirthing tools are the equally important, yet more intangible, elements, which also impact your birthing journey. These present themselves in different ways for different people, but the common thread that underpins the change many Calm Birth School students outline is a newfound confidence, self-belief in the birthing process, faith in your body and not being afraid to share those beliefs with others. The Calm Birth School's approach to hypnobirthing aims to create the most positive birth experience for you, your baby and everyone involved in your birth.

≈ Lisa's birth story ≈

Being pregnant with non-identical twins was a major shock: we discovered the girls at the 12-week scan and I will never forget the expletives that came out of my mouth. I think I cried for a week.

It soon became clear that I would have to relinquish some control to a specialist midwife (and when I say 'specialist', I mean useless) and go from giving birth in the local birthing centre in the water, to a possible induction or even a C-section in hospital. I was told this was standard protocol for twins, even low-risk babies like ours.

This wasn't how I'd imagined it! I was fed horror stories about what could happen to the twins and how small they would be. I was laughed at by a doctor when I requested a water birth, and another doctor even told me to take an epidural because, and I quote, 'Even if twin two is in the right position, I may have to put my hands inside you and pull it down the birth canal.'

From that moment, I vowed I would take back control and have the peaceful water birth that my babies and I deserved. I started The Calm Birth Course and had a one-to-one Skype call with Suzy, which gave me the confidence to contact the midwife supervisor to complain about how I was being treated. Thankfully, she completely agreed with me and we put a plan in place. I had to plan for the 'what-ifs' too, but I didn't write these down; I only wrote down the birth I wanted.

I put my consultant's nose out of joint by refusing an induction between 36 and 38 weeks, as I felt confident that my body would know what to do when the time came. I had saved The Calm Birth School's affirmations onto my phone, and practised the breathing exercises and did my Birth Rehearsal almost nightly in the bath, as well as going over the videos daily and listening to Bob Marley. I swear that Bob got me through the last few months of pregnancy.

I was given two sweeps a week apart but, despite making me surge very strongly, they came to nothing. And at 39 weeks, I reluctantly went along to my consultant's appointment where I gave in and let her book me in for an induction. I must admit I got home from the appointment and cried. I felt like my body had let me down and I couldn't believe that something I'd fought against so strongly was going to happen.

Initially, I had been determined that I wasn't going to go to the hospital, but all the surges over the past two weeks had taken their toll. I was wound up and anxious, and I felt like I was pushing my other kids – who were getting overexcited to meet their sisters – from pillar to post.

The next day, we went to the hospital for the induction. The midwife looking after me gave me a sweep before popping in the Propess.[5] She left the room and told me it often took 24 hours, even for women who had birthed before, and to have a walk and get some lunch. Within 10 minutes I was looking at my husband and telling him that the surges were strong. Looking at the monitor I was strapped to, I could see that they were peaking.

We called the midwife back and she checked me. She said I was 4cm dilated and my waters were bulging. She pulled out the Propess and said, 'Well, that's not needed then!' I was so relieved. And 40 minutes later, I was begging to get into the birthing pool.

As soon as I entered the water, I felt all the tension leave my body. I used the breathing techniques I'd practised so often and Bob Marley was on repeat. I did the breathing techniques for four hours before I began to feel the urge to bear down. At that point I grabbed the gas and air to birth our first daughter, Indy.

There was suddenly an influx of doctors dressed in scrubs with a scanning machine asking me to get out of the water to scan me. I refused. They began to scan me in the water and the doctor said, 'I can't see a head,' to which I replied, 'That's because I'm pushing it out.' Indy arrived, and we welcomed Zadie five minutes after her sister. Indy Florence weighed 7lbs and Zadie George weighed 6lbs.

Even though I doubted myself, I managed to give my girls the perfect birth. Everything just fell into place, even down to the midwife who was so encouraging. I had my placentas encapsulated[6] and the girls latched on to feed within 30 minutes of being born.

I would champion hypnobirthing all day, every day! It really helped me to draw on my inner warrior when I needed her. Now, I'm even thinking of training as a doula (see pages 68–69*) specializing in twin births, to show my fellow twin mamas that we have nothing to fear.*

Thank you, TCBS (and thank you, Instagram, for helping me to find it).

How do you create your own calm birth?

It's rare that women see other women giving birth in Western society today. Labour tends to be a secretive event that takes place hidden away in a hospital room or in your own home. This huge shift towards privacy has only come about in the last 45 years or so, but it has made an unimaginably massive difference in the way we view and experience birth. Many women will never have seen a labouring woman until they are in labour themselves.

This has left us in an interesting place. For many women, the only insight we get into one of our most significant rites of passage will come from the TV or movies, where it's often sensationalized, dramatized and presented as a torturous experience: a woman lies on her back, her face screwed up in agony, and she screams and pushes and writhes in pain until her baby comes out. This has skewed our perception of what labour is like. We go into birth fearing the worst because, let's be honest, most of the time, what we have seen in the media looks bloody horrendous. Our sanitized culture has also invited us to be squeamish and recoil from anything involving bodily fluids.

On the rare occasion that we remember we're mammals, we might look to how other warm-blooded animals give birth, in a documentary, for example. But it's easy to see an animal's four legs compared to our two and to assume that birth is going to be easier, quicker and less painful for animals because 'we humans are just built differently'.

The problem with this assumption is that it's just plain wrong. Our physical design is not the issue. If that was the case, how do you explain the thousands of women who give birth every year using hypnobirthing techniques – often without any pain relief – who recall their births as easy, comfortable and joyful? You may be tempted to write off those experiences by saying those women have higher pain thresholds, bigger pelvises, gave birth to smaller babies or any other excuse – I mean story – which springs to mind. These are, however, not the real reasons why some women's birth experiences are so empowering. There is more at work here.

While the wide-scale medicalization of birth means we are fortunate to be able to tap into world-class medical expertise in the West, it also means more women opt to give birth in hospitals rather than at home. Where women once witnessed other women giving birth many times as mothers, sisters and friends, therefore thinking of birth as normal and natural, nowadays birth is often perceived as a mysterious, scary and necessary evil, one we have to endure to get to the 'good bit' at the end. This has left us feeling less confident about our ability to give birth – a job we were made for – without medical intervention and assistance.

There is a simple reason why a woman who doesn't do the type of preparation you're about to undertake will be unlikely to give birth in the same way as any other female mammal on the planet – comfortably, calmly and without fear. This reason has nothing to do with design features. In fact, for low risk, straightforward pregnancies, the pleasant delivery of a newborn has everything to do with what's going on in your mind as you prepare for and give birth. The real issue is in the mind, not the body. Now that we know this, we can address it.

The evolution of birth

When we look at evolution, female physiology and the fact that women have ensured the survival of the human race over the last 200,000 years, it's fair to say we have done a pretty good job, right? Our ancestors somehow instinctively knew how to give birth. It would have been perceived as an everyday life event, which demystified it and stopped it from being scary. This genetic wisdom has been imprinted in your DNA and I see it as The Calm Birth School's job to provide you with the tools and techniques needed to access that information. This is knowledge that you have always had at your fingertips but which has been drowned out by the cultural noise we have lived with in Western society in recent decades. Luckily, there's no rocket science involved in getting tuned in, but we do need to get a bit scientific.

Biology and neuroscience

I'm not a scientist, but the fundamentals of hypnobirthing and how it works are based on the biology of the body and the science of the brain, otherwise known as neuroscience.

For the vast majority of women, the instinctive response to the onset of labour is physical and emotional stress. This stress response is triggered in the limbic system, which sits in the oldest part of the brain, formed around 200,000 to 250,000 years ago.

When we look back to those times, the most important aspects of human life were reproduction and survival (some argue that they still are). Language did not play a role in life; everything was based on the messages we received via our five senses of sight, touch, sound, taste and scent. These cues triggered chemical and physiological changes within the body, determining how we felt and what we needed to do, whether that meant running for our lives, fighting to the death, or freezing and pretending we were dead to protect ourselves.

The part of the brain responsible for logical, rational thinking and critical analysis, the neocortex, didn't even exist at this point. Life centred on the instinctive responses needed for survival, in order to protect ourselves from the dangers and threats of the outside world. But this ancient hardwiring is still present in us now. The limbic system continues to protect us from danger in exactly the same way today as it did hundreds of thousands of years ago. When we perceive a danger cue (whether it's real or imaginary) the limbic system, which is designed to keep us safe, wins in the fight between instinct and thinking about the best course of action logically. In short, it overrides the neocortex and tips us into fight–flight–freeze mode.

When we bring this back to the context of birth, although most women in the West are not facing life-or-death scenarios, the fear that arises when giving birth inevitably triggers the fight–flight–freeze response. In turn, this sets off a physiological reaction: stress hormones adrenaline and cortisol flood the body (don't worry, we'll go into the role they play within birth later in the book), and the automatic emotional chain of events this sets off is what can prevent us from birthing optimally in the way we were designed to do.

Part of TCBM approach to hypnobirthing is ensuring that you understand how to minimize the fight–flight–freeze response by working with the biology of your body, instead of against it. In addition, this approach will help you to start thinking and feeling differently around the expectations you have about birth. This will support you before, during and after the actual act of giving birth.

The neuroscience bit

I'm guessing you've heard about a technique called 'visualization'. This is another one of those words that people sometimes think is a bit 'out there' or New Age. However, it is to become a vital tool in your hypnobirthing journey and you'll find masses of information, evidence and case studies on it if you look at psychology and the science of sport.

Before Tiger Woods became known as 'the adulterer', he was widely acclaimed as the greatest golfer the world had ever seen – and he is an avid visualizer. You might also have heard of Jim Carrey. What about Will Smith? David Beckham? Oprah? Come on now, you must have heard of Oprah! Seriously though, all of these multimillion-dollar high achievers have spoken widely on the power of visualization. So what has this got to do with how you can give birth calmly, comfortably and more positively? It's all about your wiring.

Whether you see something in real life or simply imagine it, the combination of what you perceive and how you feel about what you perceive creates what is known as a 'neural pathway' in your brain. This pathway helps you to determine how you're going to respond to stimuli that come your way. Think of it as a blueprint that your brain calls upon when faced with a situation in which you feel threatened.

When it comes to birth, whether you have consciously thought about your forthcoming experience or not, you'll have seen many negative images and depictions of what birth is going to be like. Even before pregnancy, you may already have created a less-than-wonderful birthing blueprint and, whether it is based on fact or fiction, this can draw your body into fight–flight–freeze mode, because it triggers the suggestion that there is danger – a need for fear or concern about your ability to birth effectively and efficiently. This gets overridden if you have witnessed or been a part of a positive birth experience, when a new blueprint can be formed through exposure to new stimuli.

David Beckham, Tiger Woods, Andy Murray and many other high-achievers and performers use visualization to create the ideal blueprint of their role within the game or in the tournament before it has happened. They teach themselves how to kick or hit the ball at the perfect trajectory in their minds, many, many times. Then, when the time comes to step out onto the pitch, golf course or court, their brains instinctively know where to go because their present situation feels so familiar. This is not a new situation to them; they have already visited the scene many times through visualization. And because they are familiar with it, they don't have to deal with the same levels of adrenaline and cortisol, which come from being plonked into an unfamiliar scenario. The fight–flight–freeze response doesn't kick in, which in turn allows them to perform at the peak of their potential.

This stuff works just as effectively in a labour room, but more on how to use this technique in the next chapter.

Let's look at what else you're going to learn and take action on:

- How to manage the instinctive fight–flight–freeze response.

- How to minimize your physiological and emotional responses to stress.

- How to promote oxytocin and endorphin production (your feel-good hormones) to aid the process of birth.

- How to create the optimal environment for giving birth.

- How to ensure your birth partner can work confidently with you and your care providers.

The key element that ties everything together is learning how to relax. Sounds good, right? It is. Let's keep going; there's a lot to cover!

Chapter 2

An Introduction to Relaxation

Mindfulness and meditation are buzzwords nowadays, and have entered the mainstream in a major way. In October 2015, the UK government pledged to invest £1 million per year into training schoolteachers how to teach mindfulness to children,[7] while global organizations such as Google, PricewaterhouseCoopers and even the UK Home Office have embraced mindfulness training for their employees. The constant drip-drip feed of information on social media about mindfulness means that most of us know that learning to switch off and still the mind will benefit us, but we just can't bring ourselves to do it.

The fear of missing out on what's going on in cyberspace often robs us of invaluable time that we need to be by ourselves.[8] Believe it or not, you need to come at the top of the priority list during pregnancy, followed by your loved ones, family and friends, with the social media world at the very bottom of the list.

Taking time out for you

The first thing I invite – no, challenge – you to do during your pregnancy is to start deliberately taking time for you. In an ideal world, you'll enjoy some quiet time for 30 minutes just focusing on your breath at least once a day. That means no television, no phone, no laptop, not even a book: just time and space where you give yourself permission to let your mind have a break from thinking, analysing, planning and doing. If things pop into your mind during this time, acknowledge them, tell them you'll address them later and bring your attention back to your breath.

Some of you'll read that and think, 'Hell no! That sounds hideous!' Others will think, 'Yes, that sounds great, but back in the real world…' I get it. You're busy. However, you're also pregnant. You're growing new life inside you.

～ Tips ～

If 30 minutes' downtime feels overwhelming or unrealistic right now, start with less — say 15 minutes — and use the following ideas to help get you started:

~ Spend 15 minutes doing nothing before you get out of bed, focusing only on your breath using one of the breathing techniques you're about to learn.

~ Break the rest of your alone time down into five-minute chunks by getting unplugged three times throughout the day and focusing on nothing but regulating your breathing.

~ If you find it hard to get time alone owing to the demands of work or small children, or both, pick a quiet moment and lock yourself in the bathroom for five minutes, where you're less likely to be disturbed.

There are lots of reasons why having some downtime to yourself is good for you and your baby, so don't resist it. You're reading this book to give yourself the best possible foundation for creating a positive birth experience for you and your little bundle of fun, and the preparation starts here, oddly, with choosing to do nothing.

It's a pretty good gig and if I haven't hammered home the message enough, here follows all the reasons why:

1. Lower stress levels – this is a BIG deal

A lot of women work or live with more than their fair share of stress. When you take the time to switch off, it actively lowers the amount of the stress hormone, cortisol, in your blood. You'll appreciate this in ways you might not even notice, but trust me, low cortisol levels are a good thing. Your baby agrees, too.

2. More effective thinking

When we're stressed or angry, we become stupid. I'm being blunt but fair here. We find it difficult to think rationally or creatively when we're in a stressed-out state and are essentially useless at problem-solving. That's why TCBS breathing technique is especially useful when you're in a stressful situation at work (you'll learn how to use this technique a little later in this chapter). Ideally you would excuse yourself from the problem, go to the lavatory and do a bit of TCBS breathing, in order to create the mental space to come up with a solution. Even if you can't excuse yourself, the simple act of remaining silent and focusing on your breath will help you disengage from the frustration and be more proactive rather than reactive.

3. Pain control

There is increasing evidence to suggest that the more we can disengage from our surroundings, such as when we go into a meditative state of mind, the more our pain receptors become less sensitive.[9]

4. Induces emotional calm

Not only are you able to handle stressful situations more easily, but also you won't get riled as often, either. Stressful situations roll off you like water off a duck's back, which is good for you and your family, and amazing for your baby.

The feedback I have received from clients is that TCBS breathing techniques are not just indispensable during birth, but are also tools that you can draw on after your baby arrives. Being able to stay calm when you feel completely responsible for this new little life while simultaneously being completely out of your comfort zone is an invaluable life skill. Start putting yourself first today and embracing the downtime that is oh so good for you.

∽ The Calm Birth School breathing technique ∽

The reason TCBS breathing technique is so effective is that your out-breath is nearly twice as long as your in-breath, and this triggers your body's natural calming reflex. During labour this breathing technique will help you to maintain a deep state of calm and you'll be able to use it in between surges (contractions) to help you remain calm and focused. You'll also use it as you feel a surge coming in and once it has subsided.

Ideally, the breath is taken in and out through the nose as opposed to the mouth as this gives you more control over the flow of air. However, please don't stress if you have a cold when you're birthing. Just breathe through your mouth – it will all be OK.

How to do it

1. Breathe in deeply to the count of four through your nose.

2. As you breathe in, imagine filling your lungs right to the bottom.

3. As you breathe out, imagine sending the breath down, so it moves around your baby, down your legs and into the tips of your toes, and then into the floor before breathing out to the count of seven.

When you're learning to use this technique, you might find it helpful to place your hands on your waist, so that you can feel the rise and fall of your abdomen as you breathe deeply. It really is as simple as that.

If you find it difficult to increase your out-breath for a count of seven, simply reduce your exhalation to a number that feels comfortable, or increase the pace. As you start to feel more relaxed, you can either up your count or increase your pace. The most important thing is not to worry if it doesn't all come together immediately. The act of bringing conscious awareness to your breathing during your pregnancy will help you once you go into labour.

When to do it

Ensure you do at least one set of TCBS breathing in the morning for five minutes, five minutes at lunchtime and five minutes again in the evening. In addition, use this technique whenever you feel stressed, whether it is at work, with your partner, or getting on or off public transport – wherever and whenever. The more that you practise TCBS breathing technique, the quicker it will become your automatic response to stress.

That last point is particularly important if you labour quickly, as this breathing technique will get you into the birthing zone quickly and easily.

If you tend to take life in your stride with very little stress, this doesn't make you exempt from practising this breathing technique. It's just as important that you carve out some practice time for your TCBS breathing, too.

As you get better and better at using TCBS breathing technique, you'll notice how the skill of being able to relax on demand – and eventually very, very deeply – becomes both instinctive and invaluable to you. And by the time you reach your labour day, you'll be able to relax and enjoy (yes, I said enjoy) the process, as your ability to release stress and tension from your body in everyday life will help you to learn to trust that your body will be able to do this for you during your labour when you need it to the most.

Hypnosis/trance

In many cases, people feel weird about hypnosis (which is often referred to as trance, and I will use the two terms interchangeably). The sense of weirdness often comes from what they have seen on stage and TV shows. While this type of hypnosis often makes for great entertainment, it is completely misleading and incredibly frustrating for anyone who uses hypnosis in a therapeutic setting, because what we see on TV or onstage is an illusion: the illusion of mind control.

Let me explain what I mean. The first thing anyone working in hypnotherapy is taught is that all hypnosis is self-hypnosis. This means that you can't make anyone do anything that doesn't sit comfortably with them. When we watch a stage hypnotist, what we're seeing is an audience made up of two kinds of people: those who are there to be entertained and those who are there to be the entertainment. We'll see a large group of people invited up onto the stage and they'll be directed to participate in 'entertaining' things that usually involve them looking a bit silly. As the show goes on, the suggestions that the stage hypnotist uses will become increasingly outrageous.

Every single person up there will have their own internal barometer that tells them when enough is enough. As each reaches that point, you'll notice them choosing not to participate on the stage, and at that point the hypnotist will tap them on the shoulder and ask them to sit down. Finally, you're left with the 'star' of the show – the person who feels the least inhibited in that environment, doing things that under normal circumstances they wouldn't want your grandmother to watch. At no point can anyone be forced to do anything they do not want to do, and the only reason that they are clucking like a chicken or barking like a dog on demand is because they want to.

I hope that sets the record straight once and for all: hypnosis has nothing to do with mind control!

So if hypnosis isn't mind control, what is it?

Hypnosis is a natural state of consciousness we drift in and out of whenever we have a narrowed focus of attention, or are deeply relaxed in passive or active activity. It's a state that we all drift in and out of at least 10–12 times every single day. The times that you'll be most familiar with are when you're just drifting off to sleep or when you're just waking up.

Other examples of natural states of trance include:

- Daydreaming

- Running

- Dancing

- Sex

- Drawing, colouring or any other creative pursuit, e.g. writing, pottery, sewing, etc.

- Driving a familiar route, e.g. to work or to the local shops

- Watching a great movie

- Listening to someone tell a story

We enter trance whenever we experience a narrowed focus of attention, becoming deeply absorbed in an activity that we are very familiar with or are extremely relaxed doing, the logical part of our mind becomes more easily sidetracked. Whenever the brain is not required to critically analyse everything we are doing, a gap in our thinking is created, and this leaves us more open to suggestion. This state of suggestibility can't occur when we are learning a new skill or taking part in an activity that requires us to be alert and aware.

This is really important to remember if you find yourself listening intently to someone else's birth story, particularly if it is about a terrible experience. The more involved and engaged you become with the story, the more open to suggestion you become. In this type of scenario your brain stops critically acknowledging that the story you're hearing is only the specific storyteller's experience, and without realizing it you start to file the information under the 'What Is True About Birth' file in your brain's hard drive. As the suggestion that birth is an awful and scary experience gets implanted in your brain through listening to this tale, it becomes your point of reference when you think about birth and, most importantly, when you go into labour.

Therefore, it is imperative that you consciously avoid listening to negative birth stories at any opportunity – even if that means removing yourself from the conversation. However, the upside of this is pretty great. As the fact that you're more open to suggestion and to storing information in your brain's database when you feel calm, relaxed or focused is exactly what we capitalize on when it comes to your hypnobirthing journey. This is the exact reason why hypnosis is so great for helping you to retrain your mind about how to think and feel about birth. It can be used just as effectively to reinforce positive suggestions, feelings and associations you have about your upcoming labour and birth.

So let's bust some hypnosis myths:

- **You can get stuck in hypnosis:** You can no more get stuck in hypnosis than you can get stuck in sleep!

- **Hypnosis is harmful:** The only time hypnosis can be harmful is when a person is in a natural state of trance – listening uncritically to someone else's suggestions about an event, such as in the example given above – and listens to someone else's negative birth story without acknowledging that the experience being recounted does not necessarily have any bearing on the way that they will give birth.

- **Hypnosis is mind control:** Well, we've already busted that one! Remember: all hypnosis is self-hypnosis.

The brain

While we're not going to dive deep into the truly fascinating world of neuroscience, understanding the basics of how your brain works will help you to get clear on why going into a state of trance, self-hypnosis or meditation by yourself is so great for birth.

I'd like you to imagine the brain being like an iceberg. The tip of the iceberg is where all of our logical, rational thinking takes place, but the subconscious, which is effectively all the stuff going on beneath the surface, is where most of the action is.

The subconscious mind runs our autonomic nervous system and is responsible for the beating of our hearts, the blinking of our eyes and anything we do without consciously thinking about it. It is also the home of our emotions, our imagination and where our gut feeling, intuition or 'sixth sense' originates.

The driving force behind everything our subconscious does is ultimately designed to keep us safe. The way it does this is by keeping records or memories – both visually and emotionally – of everything we have ever seen or been a part of, and categorizing them crudely as either 'safe' or 'unsafe'. If an event takes place for which we don't have a frame of reference (in other words, a brand new experience), the subconscious will look for the most similar thing we have experienced and then trigger a set of physical and emotional responses in line with that reference point.

An example of this might be someone witnessing another person getting hit by a car. If they haven't been able to process the incident logically, the memory of seeing the accident will trigger a chain of physiological and emotional responses, which might result in the subconscious telling the person instinctively, 'It's no longer safe to cross the road.'

This is what happens so frequently when it comes to birth. Unless you have experienced a difficult birth, most women will have only witnessed challenging births on TV or heard other people's stories. Birth stories, and more specifically negative birth stories, that are shared with pregnant women are interesting to me. It's clear from the perspective of the media that drama sells. To create stories around women who have calm, positive, relaxed births is simply not as exciting as women who punch, kick, swear and create the 'on the edge' illusion of 'will she actually be able to go through with this?'.

But when it comes to us as sisters, real women recounting our tales, I don't think it's a lot more complex than just assuming we also just love the drama. While I know from surveying Calm Birth School fans and students that those with positive birth stories feel like they stand alone in a sea of negative ones, there's a sense of not wanting to be 'that' woman who recounts taking labour and birth in her stride. So instead of speaking up, she just keeps her head down.

I also think there is a sense of camaraderie around shared pain. And unfortunately for many women who haven't experienced a positive birth, sharing their story acts as a warning and sometimes feels like part of their healing. I think if you can relate to this, one of the best things you can do is to speak with a trained professional to get a comprehensive debrief on your birth story. Asking your hospital or birth centre to talk through your birth notes can often be a great step towards healing. And for those women who recognize that they have experienced some aspect of trauma, working with a doula, a hypnotherapist or even a counsellor specifically trained in helping women who have experienced birth trauma can be the missing link to moving beyond painful, negative memories.

Without the appropriate counselling, hearing – or watching – other people's birth stories may cause a pregnant woman to experience a negative charge of emotion, which she then files into her brain's 'unsafe' database. Should she then be faced with a similar situation (i.e. going into labour), the subconscious mind may try to protect her from having the baby by releasing hormones that activate the survival mode – the fight–flight–freeze response that we discussed in Chapter 1 (*see page 8*). In turn this can impede her ability to birth calmly, comfortably and efficiently, as her subconscious attempts to protect/stop the 'unsafe' event from taking place.

⮞ Tip ⮜

If someone tries to tell you their birthing horror story you have my express permission (actually, this is an order!) to get friendly with your inner child, put your hands over your ears and sing the 'La, la, la, I'm not listening' song. If that isn't quite appropriate, stop the conversation politely and firmly tell the storyteller you would prefer to catch up with her birthing story after you've had your own experience. You'll feel great once you have set a clear boundary and you won't be adding any negative stories to your database. If you find setting boundaries challenging, this is a great opportunity for you to start to tap into that protective maternal energy and use it to take care of yourself and your baby.

The part of the brain responsible for activating the fight-flight–freeze response sits in the oldest part of the human brain, the amygdala, which doesn't understand language or logic. It responds to situations based on imagery and feeling: what we see, even if only in our imaginations, and how we feel at a particular time.

It is interesting that the body's hormonal and physical responses are incredibly similar whether we are seeing an event for real, imagining it or dreaming about it. This is why your skin feels damp from sweat and you experience an increased heart rate when having a nightmare. Even though you're dreaming, the brain can't tell the difference between what is real and what is imagined. It responds to what you're seeing and feeling in your dream in the same way as though you were having the experience in real life. Let me reiterate: the brain can't distinguish what is real from what is unreal. If you perceive an event to be 'true', either through your eyes or through your imagination, the brain will process that event as though it is really happening and your body will respond accordingly.

In order to change the physiological chain of events that occur when you have decided – either consciously or subconsciously – that birth is unsafe, you literally have to retrain your brain to expect something different.

A simple process to help you retrain your brain is imagining or visualizing your ideal birth scenario and regularly listening to new audio input, such as MP3 meditations. This simple process is so powerful because it enables you to begin to create a whole database of new 'memories' about how positive your birth is going to be, meaning you can change which category your 'birth' files are stored in, moving them from being labelled as unsafe and scary to being safe and positive.

I know if you're a first-time mother you might be thinking, 'Yeah, but I've got no idea where to start; I've never done this before.' However, a great resource for helping you with this is YouTube. Type 'hypnobirth' into the search bar and you'll be presented with some great examples of women giving birth calmly and positively. My personal favourite is 'Daisy's Hypnobirth Homebirth' with Mama Susy O'Dare. I love how calm, confident and in control Susy is, and how gently she welcomes baby Daisy into the world. Look for new YouTube inspiration once a week and you'll be off to a flying start.

Watching positive hypnobirthing videos regularly serves two purposes:

1. It gives you a real and tangible frame of reference that you can use to start building your own new blueprint for what birth can be like.

2. You can start to think about the different stages of labour and birth you'll experience and, if you particularly like what you see, mentally superimpose yourself into the films. Or think about how you would like to do things differently.

∾ Create a positive birth story ∾

In the following practice, I want you to start thinking about exactly how you'd like to give birth. But when you think or write down your birth story use the past tense, as though it has already happened.

Why should you write your story out in the past tense? A simple way to think about how the brain works is that it is a problem solver. If we ask ourselves a question, our brain will always look for the answer. By writing out your birth story in the past tense, you close the 'question–answer' loop in advance. I'll explain more about why this is important to you and your baby when we discuss the role of visualization during your birth in a later chapter.

How to do it

Use the following questions to help you to start thinking about what you want to see, hear and feel during labour:

- Your blueprint or story may start with how you felt when you woke up in the morning or at night with those first few cramps. How did you feel? Calm? Excited? Happy?

- What was the conversation like between you and your birth partner? Were they there with you? Or did you have to call them to let them know that you thought things had started?

- What did you do next?

- Did you go for a walk?

- Prepare some food?

- Did you watch TV?

- How did your body feel as your labour progressed?

- What rooms/buildings were you in?

- What did the room look like?

- Who was there with you?

- What did your body feel like when baby's head began to emerge?

If you find this too difficult to write out now, don't worry. Bookmark this page and write out your story once you have completed the book, if writing is useful. Or just start daydreaming/visualizing about the day you'll meet your precious little one.

When to do it

Once you have your ideal birth scenario in mind, daydream/ visualize this story as frequently as possible; you can't think about it too often. When you're in the car, any 'dead time' you get at home, or when you're travelling to work, just start thinking and feeling into how you're going to be. The sooner you start creating this positive visualization, the sooner you'll start creating those new neural pathways for your brain to follow once the time is right.

You might have read through the above practice and now be thinking, but I'm not a visual person.

Don't worry, I'm not visual at all, which is why I have you covered.

Perhaps you're more of an auditory person who loves to listen. In this case, working your way through the MP3s that accompany this book, you'll really be able to embed the teachings by listening as you relax and even sleep. Although you might not have the visual pictures in your mind, the consistent repetition of the same positive approach to birth will create new blueprints for your brain to refer to when you go into labour.

If you're like me and what they call a 'kinaesthetic learner', you'll probably learn best when you can move around and feel. So the way you can use this to your advantage when creating your 'visualization' is to think about how you want to feel during each stage. Imagine how relaxed the muscles in your face will feel if your surges intensify. Imagine your body feeling completely tension-free, like a rag doll. Maybe you'd even see yourself laughing and crying with joy once you are finally holding your baby. It's your birth, how do you want to feel? You decide.

✑ Emiliana's birth story ✑

After a manic few weeks of moving house a few days before my due date, I decided to treat myself to a pregnancy massage. We told Bump that 5 September would be the optimal date she could come: her dad, James, would be home from work and it would give us enough time to settle into the new house (not unpack, just settle!).

The massage left me feeling calm and wonderful, and James said I looked like a new person when I got home. In the evening we had friends over for an easy pizza dinner, and I'm sure it was feeling so relaxed and happy that got things moving...

I didn't feel any different when I went to bed, but the surges started at 6 a.m. and felt like mild period pains. I knew they were surges because when I timed them they were evenly spaced.

I woke James about 45 minutes later and he kicked into action, making the living room all lovely with candles and incense. I ate some cereal and drank tea, thinking of all TCBS stories I'd read in which things took a little while from here. As we were hoping for a home birth, we decided to call the midwife at around 8 a.m. She said she'd be at least another 30 minutes, so I carried on welcoming the surges and laying out the collection of birthing tools (jelly babies, water, straws, paracetamol and so on) that I might need, as well as getting the TENS machine going.

When the midwife arrived, she checked my progress and I felt a bit disappointed when she said I was only 3cm dilated but immediately Suzy's voice popped into my head saying, 'It's only a snapshot in time!' She was totally right because 90 minutes later, after a bath and more powerful surges, I was 7cm.

After some funky rocking movements standing-up with James holding on to me from behind (we are still laughing at how ridiculous we must have looked, but it seemed like the only way to ease the discomfort!), my waters released. I got into the pool just before 11 a.m. and that's when the midwife noted that I was in active labour. A second midwife arrived and the show really got on the road!

I tried a few different positions in the pool over the next three hours: head against the side taking in gas and air, and then back around, sat on the inflatable stool with legs up and my hands grasping the back of my legs. James was always right behind me, whispering words of encouragement and handing me the gas and air when I needed it.

During the last 30 minutes, the midwife respected my TCBS notes (she did the whole time, but I particularly noticed it here), and talked about breathing the baby down and what it would feel like when she was finally in my arms. James did the same and that positive visualization, together with allowing my body to take over, made all the difference in those last three hours. When Mollie's head started to crown, I didn't experience any tearing. Again, I put this down to letting my body and gravity do its thing and breathed rather than pushed when I needed to.

But it took a lot out of me and by the time her head was out, I had lost all my energy and heard the midwife talking about Mollie's heart rate creeping up, which spurred me on to give the next surge a big helping hand and focus on pushing. We had to wait three minutes for the next contraction and James said it was the longest three minutes of his life! But I talked to Mollie, saying we had to do this one together as the team we were.

When that surge finally came, she shot out and the cord, being a little shorter than normal, snapped. Mollie's end clamped itself naturally (amazing!), but I lost 400ml of blood, which I didn't notice in the euphoria of holding our newborn baby. I had wanted a physiological third stage, but the placenta wasn't playing along and with the blood loss the midwives advised a managed third stage, as they call it (see also page 179). By that point, I was so happy with the way everything had gone that I didn't mind some intervention and just wanted to spend time with our little baby as a new family.

Giving birth to Mollie was a wonderful, joyful experience and I put that down to TCBS techniques, a positive mindset, being at home and the birthing pool. It was an amazing start to our journey into parenthood – thank you.

Chapter 3

The Biology Bit

hope you're now beginning to see how hypnobirthing is based in science – fist bumps all round. However, if you're still feeling sceptical, this chapter will explain how staying calm and relaxed can have a positive physiological and emotional impact on your body and your ability to give birth the way you designed.

The uterus and what happens in labour

No one knows why women go into labour precisely when they do. If I knew the formula to that one I'd be a very rich lady indeed! What we do know is that the body generates a huge surge of the hormone oxytocin, which stimulates the muscles of the uterus and kick-starts labour.

There are two types of muscle in the human body: voluntary, such as those in the arms and legs, which we can control consciously – and involuntary, such as the heart and uterus, which we can't.

The two layers of uterine muscles, one horizontal and one vertical, work together as a pair. They operate in the same way as any other muscle in the body, but you can't consciously move or control them. When you go into labour, the vertical layer of muscles moves down over the horizontal layer and begins to pull the horizontal layer upwards. When a woman is calm and relaxed, this movement is smooth, the muscles working together in harmony. The upwards motion causes the neck of the cervix to thin and open so that the baby can move down the birth path easily, without stress.

You can see what this looks like here:

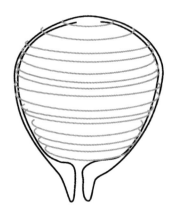

Figure 1: Inner layer of uterine muscle

The inner layer of the uterus is made up of circular horizontal muscles, which are located in the lower area of the uterus, with the thickest ones situated just above the opening or neck of the uterus (the cervix). For the baby to move easily down into the birth path, these thicker muscles must be drawn up and back.

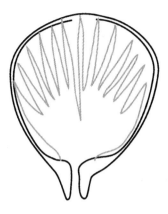

Figure 2: Outer layer of uterine muscle

The outer layer of the uterus is made up of stronger vertical muscles. These muscles go up the back and over the top of the uterus, drawing up the relaxed circular muscles of the inner layer.

Figure 3: The uterus in surge

When the birthing mother is in a state of relaxation, the two sets of muscles work in harmony in a wave-like motion. The vertical muscles draw up, flex and expel, and the inner circular muscles relax, open and draw back. Birthing then takes place smoothly and easily. Remember what I said earlier: your body is designed to give birth.

When you have a surge during labour, you may notice your tummy gets very tight and you can see things noticeably lifting upwards, as shown by the dashed line in Figure 3 on the previous page. The uterus lifts as the muscles work during a surge and then return to their normal position.

When a woman is tense or scared, the involuntary muscles of the uterus are prevented from working together in harmony. As the horizontal muscles tense up, the vertical muscles attempt to pull up these rigid, inflexible horizontal muscles. This state of muscular tension, along with your baby's head putting pressure on a cervix that is not thinning or opening because of lack of movement, causes the pain that many women experience during labour. This experience confirms a woman's initial fear that labour is indeed painful, which keeps her in a tense, stressed state.

Author of *What Size Is Your Brain?* Veronique Strohbach says: 'For every thought we have there is a corresponding physical and chemical reaction in the body.' Assuming we have healthy mothers and babies, this quote becomes the cornerstone of our understanding around how and why the uterus either works efficiently or becomes tense and inefficient when we are birthing.

In the previous chapter, I explained how the brain's database of memories sends messages to the body about what is happening (*see page 22*). For some women, another person's story will flash through their mind during labour, sending out a red alert that something is wrong. For others, something or someone within the birthing environment may trigger feelings of unease or being unsafe, which in turn puts the body into survival mode, as described in Chapter 1 (*see page 10*).

One of the body's key chemical responses to a conscious or subconscious thought of danger or threat is the release of adrenaline. Adrenaline isn't a friend of labour. When we go into survival mode the physical response of the body is to divert blood and oxygen away from the uterus, because in evolutionary terms, the uterus would not have kept us safe from whatever the danger – real or perceived – was. As much blood and oxygen as possible is sent to our extremities because of their ability to help us to fight for our lives (with our hands) or flee for our lives (with our feet). The uterus is understandably deemed useless and, as such, it is starved of the blood and oxygen it needs to function, thus becoming less and less efficient.

This is great if you're about to be attacked by a lion, but when it comes to labour, this is not a good scenario: the lack of blood and oxygen available not only slows the process down, but also makes it more painful. The pain then reinforces the initial thoughts of fear, confirming there really is something to worry about. As labour slows and the birthing mother becomes tired, she is much more likely to need intervention as her body, and baby, becomes less able to cope with the impact of the stress response.

In other words: fear equals tension equals pain.

> **~ Tip ~**
>
> Whenever you're feeling stressed, but particularly in the run-up to and during labour, one of the best things you can do to counteract the fight-flight-freeze response is to bring your attention to your breath and focus. TCBS breathing technique you learned in Chapter 2 (see pages 16–18) will help to neutralize the stress response and create a release of endorphins.

Why endorphins are your new best friend

Endorphins are amazing for birth and a great friend of labour. They act in a similar way to opiates in the body, and are said to be up to 200 times more effective than morphine when it comes to pain relief – which is why you ought to love them. The more things you can do to relax and feel good during the early stages of your labour, the more of those happy hormones you'll bank, helping you to feel more at ease, more in control and more comfortable.

The other hormone I mentioned earlier, and one you need for an efficient labour, is oxytocin. Oxytocin is 'the hormone of love', a chemical we release into our body whenever we feel love for something or someone. When you make love to your partner, your body courses with oxytocin. (It's the reason many of you are reading this book with your lovely bump!) Oxytocin is the hormone that gets labour started. When you're able to stay feeling calm and at ease, it's also the hormone that causes some women to feel euphoric and even joyful while birthing. A woman receives the biggest peak of oxytocin at the end of labour. This happens in between birthing baby and expelling the placenta (or afterbirth as some people like to refer to it).

Oxytocin is the reason why, no matter how moulded, squished or messy baby's head will appear when they are born, many women look down into their newborn's eyes and fall totally and utterly in love at first sight.

This is part of nature's very clever evolutionary tool. It makes us want to stay close and look after our babies, which ensures the survival of the human race. The physiology of birth remains pretty much the same as it was over 200,000 years ago when Homo sapiens first appeared. Not only did nature get the third stage right, birthing the placenta, but it also got the first and second stages, early labour and the birthing phase, pretty spot on, too. Healthy women needed to be able to give birth as efficiently and as enjoyably as possible, because when we revisit what early human society looked like, there were no doctors, midwives or doulas (in the form that we know them, at least) at all. Yet women still gave birth again and again many times over, until we found ourselves here in the 21st century.

In short, we women were designed to do this job. As you release resistance in the lead-up to giving birth, and deepen your trust in the knowledge that your body and baby know what to do, when you reach your labour day, you really can let go and follow whichever direction your baby decides to take you in.

Pain

The time has come to put something on the table: we need to talk about pain. By now you should have a very clear understanding of the science behind why some women birth really comfortably, while others do not. As I mentioned earlier, there is a relationship between the kind of thoughts you think, the amount of fear you feel, how tense your body becomes because of that fear, and how much pain you experience. Once again: **Fear = Tension = Pain.**

Does this automatically mean that if you're fearless, you won't feel any pain? No: there are no hard-and-fast rules about how you'll experience the sensations in your body when you're giving birth.

Some of you reading this book will birth in complete and utter comfort, feeling relaxed and open, without resistance and able to birth your baby while feeling sensations but not pain. Others will feel sensations that might previously have been described as pain, but by preparing for your birth and mastering the techniques in this book, you'll feel that the experience is totally manageable. You'll find that you can distance yourself from the experience by using your TCBS techniques to keep you in the zone. Or you might also get to the other side of your labour and say, 'It was bloody painful, but I rocked it!'

Every woman's experience is subjective and not something you can measure your own birth against. Whether you feel pain or not, whether you roar like a lioness or focus all of your energy inwards, giving birth will be your own unique journey. If you're birthing without fear, and feel able to surrender to whatever it is your body calls for you to do in the moment, you'll have experienced a positive hypnobirth.

And be mindful that we all experience pain differently, depending on how we are feeling emotionally and what our brain decides is the most important focus in any given moment. If you have bad period pains but a lot of work to do, you're more likely to power through things, focusing on the task at hand rather than worrying about how your abdomen feels. But the minute you finish your chores and sit down, the pain suddenly hits you. This is because you no longer have the distraction of your to-do list to focus on. This is why your breathing techniques, visualizations and, for some people, their MP3s play such an important role in their births, as they become a distraction from the sensations of labour.

The key marker for women who do experience intense or painful births, but who also have the tools of hypnobirthing in their back pocket is that they understand that whatever they are experiencing is totally normal and nothing to be fearful of. This allows them to redirect their focus and choose to experience the physical sensations in a different way.

～ Tip ～

One of my favourite affirmations, reinforcing the idea we can choose to feel the sensations of birth in whatever way we desire is:

'Each surge brings me one step closer to meeting my baby.'

You can see how thinking in that way makes the sensations far more appealing and something to be welcomed from this perspective.

If, when you're birthing, you think about what a nightmare everything is, and when you are going to feel the next surge and for how long, the whole experience becomes tiring, draining and intense. So from here on in, if you haven't been doing so already, rather than focus on how painful labour is going to be, be curious about which techniques are going to help you to remain comfortable – this will be much more helpful.

I also recommend that you explicitly ask anyone involved in your birth to talk about your comfort level, as opposed to how much pain you're in – more on this in Chapter 7, when we'll also address communicating with your care provider. Remember that when you're sat in a state of relaxed consciousness, you'll be far more open to suggestion; you don't want a well-meaning midwife making an indirect suggestion that you need assistance because you could be coping more effectively with the pain, and for you to start analysing that, taking her suggestions on board.

A woman who is able to look at the sensations within her body as part of the normal process of birth, feeling at ease with what she's experiencing and knowing she doesn't need to fight against it finds the whole process much easier. She becomes able to experience the intensity of birth without fear, knowing that the sensations she is feeling are normal and natural, and reminding herself that all she has to do is allow her body to go with it. She releases all resistance. The way she experiences that intensity or pain becomes a wave that she is able to ride, rather than something she is trying to fight against, as Karis' story below demonstrates.

➳ Karis' birth story ➳

At 41+3 weeks I relented and agreed to a sweep, having refused one the week before. We made the decision, with the help of TCBS, that a sweep was really OK and may help things move on a bit. My parents had come over from the UK for one of the guess dates and I was feeling the need to get a wriggle on (not that there was any pressure from them; it was my perception).

We were getting the endorphins going left, right and centre. I was still riding my bike and going for walks; my parents even went out, saying to us, 'Go on, have vigorous sex!' (Not what you want to hear from your church-going mum.) I even went for a jog!

The sweep didn't work but my blood pressure was high, so the midwife told us to call the hospital. After being monitored twice previously during the final month, I was told it wasn't high enough to worry about, so off we went back home. After supper that evening I felt strange and knew something wasn't right. I took my blood pressure again and it was much higher than before, so back we went to see the midwife and this time we were sent straight to hospital.

My plan to labour at home and have a water birth was off the table, as after 41 weeks you're no longer under midwife care. I was monitored overnight. My blood pressure remained high but stable. My mum has had a brain haemorrhage in the past so we didn't want to ignore the high BP, and I was happy that baby and I were being monitored.

The next morning, I was given a Foley Catheter to encourage my cervix to move and start to dilate. Was it invasive? Possibly, but I should say that having read the statistics I knew that my baby was not coming out naturally, and my TCBS breathing and affirmations helped me through the procedure. I was also confident that the birth of our baby would be great whatever happened. After living with polycystic ovary syndrome (PCOS), we didn't think we would even be able to have a baby, so the end result was going to be worth it, regardless of the process.

That night Matt went home for some sleep (lucky him) and then the surges started. The catheter was doing something, at least. By 2 a.m. a giant Dutch midwife with the longest fingers ever came to see me and check on my progress. My cervix was still being backwards in coming forwards but she could at least touch the membranes. Sweep done, catheter removed: time for me to get some rest.

The next morning, the doctors decided an induction was required, as my BP wasn't improving and they wanted baby out.

Once in the birthing suite, I was hooked up to machines to monitor my BP and the baby's heart rate but I could still move around a bit, sitting, standing, squatting. Then labour began in earnest. I wasn't prepared for the strength of the hormone-induced surges; neither was I aware that there would be no let-up or break between them. My first thought, 'What the heck?!' was followed by Suzy's voice in my head: 'Breathe, breathe, breathe.' I had affirmations running through my mind the whole time, and at one point Metallica's Enter the Sandman *started blaring out, with Matt beaming at me. We were having such a good time. No, it wasn't what we'd planned, but we were going to meet our baby boy or girl by the end of the day!*

At 6cm dilated I requested an epidural, thinking I would get a reprieve. I'm a potty-mouthed swearer, but through the entire process the one and only time I decided to swear was when the anaesthetist gave me the pre-epidural prick. I shouldn't really have called him a MoFo. First, he and his assistant were both hot, and second I would be seeing him later. Oops!

The surges came thick and fast after that. I thought the epidural would help or stop the discomfort, but it just changed it. I got onto the bed and Matt held one of my hands palm open while I used my other hand to stroke the bed bar up and down in time with my counting. Anyone in Room 11 will have the shiniest bar on the bed! It was working; we were doing this and loving it.

The midwife came back into the room to examine me: 8cm, whoop whoop! However, baby's heart rate was not recovering well. They did a blood gas test on baby's head, and the results were right on the cusp of their preferred readings. The midwife and nurse started discussing something in Dutch. I said, 'You're preparing for a C-section, aren't you?' They told me that they always like to be prepared. The second test was done and the tempo shifted gears; it was clear that they wanted to get baby out. They knew what my birth wish had been and we were so far away from it that they didn't want to tell me. I was absolutely fine with the change of plan; I even said, 'Brilliant, let's go.' For us, it was all about the journey to meet our Boo Boos, who arrived at 8:12 p.m. on Friday 15 May covered in poop: a beautiful baby boy weighing 7lbs 8oz, not the behemoth we had been told to expect.

William had arrived, our miracle – the baby we thought we would never have – on a day that was so happy, enjoyable and mind-blowing owing to the work we put in with TCBS. I want to say to any woman preparing to give birth that even if your birth wishes don't happen, you can be in control of yourself and your reactions to the unfolding situation.

～ Wave breathing ～

Use wave breathing when you're experiencing a wave or surge.

How to do it

The central idea of wave breathing is to keep both the inhalation and the exhalation even. Breathe in through your nose for the count of seven and out through your mouth for the count of seven.

The role of this breath is to work with the upwards motion of the uterus as it rises (*see Figures 1-3, pages 34-35*), and then to send your breath down to your baby and your womb, while relaxing. Simple!

However, don't be fooled. To move instinctively into that space of deep breathing and relaxation when you experience a wave, you need to have practised it so often that it is second nature.

Please do not worry if you're unable to keep the breath even for a count of seven to start with. Work with whatever feels most comfortable for you. Perhaps start off by counting to four and once that feels good extend it to five. The main point is to become comfortable slowing your breathing down and taking control of the flow. This will help you immeasurably during labour and birth.

When to do it

Aim to practise the wave breathing technique every morning for five minutes. If that means setting your alarm five minutes earlier – do it. It's such a good way to start your day and will leave you feeling great as well as preparing you for your labour day, when you'll be using it during each surge you experience.

Chapter 4

Taking Control:
Do the Work

As you can probably tell by now, I like to tell it as it is – no messing around. I've explained that hypnobirthing isn't a magic wand, but it is one of the best things you can do to give yourself a great foundation for making your birth as easy, comfortable and positive as possible. (Yes, I will keep repeating this over and over. We have to make sure the old fear-based message is well and truly erased!)

By the time you have finished reading this book, you'll have all the tools that I have shared with thousands of women around the world who have created amazing births, the kind they're happy to share with anyone who'll listen! However, reading the book and listening to the MP3s isn't all there is for you to do. You have got some work to do in order to experience the type of positive birth you've been reading about. Actually, make that a lot of work. Yes, you get points for showing up and reading, but that is just the beginning. The time has come for us to make an agreement. This may or may not go against all of your sensibilities, but it doesn't matter. Wherever you are right now, I ask that you suspend any and all beliefs caused by any doubt about your ability to create a really positive birth.

What is a belief?

Essentially, a belief is a thought that we have repeated to ourselves so many times before that we have categorized it as 'real'. By repeating your thoughts about your ability to create a calm and positive birth experience tens, hundreds or even thousands of times, you start to form a new belief for your subconscious to follow. This is a hugely important part of your preparation for your birth. Helping you to look forward to your birth feeling even more confident as you mentally rehearse your birthing experience and your ability to navigate all or any of the situations that may be presented to you with ease.

Remember the way the brain works? When you believe something to be true – even if it's not actually true (in this case, that birth is painful and difficult and it's only fun once you get to the end bit), the subconscious will do its very best to prove you right and prevent you from experiencing this unsafe event. In practice this can take the shape of slowing down or even stopping your labour, as it tries to stop you from giving birth in order to keep you safe. This is obviously not what you want to happen. If you're in labour, you want your baby to come out, not to stay in, and delaying this process usually results in more discomfort and medical intervention that might not otherwise be necessary.

We embed new ways of thinking and responding to situations through constant repetition of new actions or thoughts. Think back to when you learned to drive; when you first started learning, you didn't think there would ever be a time when you could have people talking to you in the back seat, while having the music on, while navigating a new city. Every ounce of concentration had to go into using your mirror-signal-manoeuvre process, and you had to practise again and again (and again).

If you don't know how to drive, perhaps you remember learning to ride a bike; it took a lot of practice before it became instinctive enough for you to ride with ease. The only way we can change our subconscious patterns of belief is to know that there is an alternative to our ingrained thought patterns.

The following exercise will help you to rewrite any unhelpful beliefs you might have about giving birth.

∾ Using affirmations to change beliefs ∾

One of the easiest ways to change your thoughts and beliefs is to use affirmations to choose consciously the things you want to say to yourself – to change your internal dialogue. Affirmations are simple, powerful, positive statements that you repeat often so that they become your new narrative to how you're going to approach and experience your birth.

How to do it

Write out the following and then say it aloud, as often as you can:

- I know I can enjoy an amazing and positive birth. I will immerse myself in positive birth stories.

- I will imagine my positive birth with feeling. I will do my daily breathing exercises.

- This is important to me because [*complete this sentence*].

49

And once you have completed that exercise, I would like you to do the same with the following sentences:

- I am creating an amazing and positive birth.

- I am immersing myself in positive birth stories. I am imagining my positive birth with feeling.

- I am doing my daily breathing exercises.

- This is important to me because [*complete this sentence*].

Changing the sentences from 'I will' to 'I am' takes things out of the future into the present tense – giving you no excuses to hide and letting your brain know that this is happening.

When to do it

Don't do this exercise later: do it now, even if it's on the back of an envelope. Don't think it in your head. Write it down and say it out loud.

It doesn't matter if you feel silly saying these words. What's important is that you're fully invested and believe in your body and your baby's ability to do the job they were designed to do. If repeating some affirmations out loud helps you with this, then it's 20 seconds well spent.

If you're already feeling confident, relaxed and committed, this is a worthwhile exercise in reminding yourself why it's important to create a positive experience. It's a win–win either way. If, however, you're feeling anxious and awkward, now is the time to leave Ms Cynical at the door.

Embedding your new thought patterns

You can turbocharge the process of embedding these new thought patterns in your mind by adding emotion. Imagine how you're going to FEEL when you look your baby in the eyes for the first time. Imagine how you're going to FEEL as those first proper surges kick in, and what it will feel like to have your birth partner observing you in all your strength as you calmly and confidently guide your baby out into the world. Play around with words, sentences and images if you can – just like I described with the visualization exercises – and get feeling.

So how can we help you achieve this? First things first: if you haven't yet said the above affirmations out loud, stop reading right now and recite them. This simple act is symbolic: you're sticking your marker in the sand and declaring, 'I'm taking control now.' As you move through the course you'll also learn how to let go of everything you can't control.

Here are the statements again. If you read them before, here's another opportunity to do it again and feel it:

- I am creating an amazing and positive birth.

- I am immersing myself in positive birth stories. I am imagining my positive birth with feeling.

- I am doing my daily breathing exercises.

- This is important to me because [*complete this sentence*].

Once you've declared this – and meant it – you might like to start listening to TCBS course of daily affirmations. There are written versions available for free at www.thecalmbirthschool.com/bookbonuses or check out my daily prompts on TCBS Facebook page. Start reading and reciting them every day.

If you're feeling sceptical of the 'war cry' above, I strongly suggest that you make time to write out and say the above 'I am' affirmations every day at least 10 times, before you listen to the recorded affirmations. Continue to practise the affirmations daily until you really believe them, then you'll be able to dive straight into the recorded affirmations without any preparation.

For some of you the words will resonate and hit home right away; for others, they won't feel real until the day you go into labour. Neither is better or worse. All you have to do is commit to getting there.

Once you've have listened to the recorded affirmations, decide which three or four you love the most. Write them down and post them around the house, on places like the bathroom mirror or the back of the front door, so you're continuously reminded of the calm and positive birth you're creating.

Dealing with worry and anxiety

Pregnancy and the thought of birth can take some women by surprise. We can be so confident in other areas of life, but when it comes to our pregnancy bump we get 'worryitis'. If you're already prone to worrying, the unknowns of pregnancy can simply add to your current anxiety. This is not bad or wrong and I've seen it many, many times before, so if you fall into this category, please do not beat yourself up for it but do use the exercise below.

~ Release and let go ~

The following technique will help you to put those worrisome thoughts to bed, so that you feel as chilled, relaxed and happy as possible, instead of tense, uptight and nervous. And the great news is that because worrying is simply a habit you have learned over the years, anything that can be learned can also be unlearned. Sometimes it takes a bit of practice, particularly in the beginning, when it often feels easier to do what you have always done, which in this case is letting yourself worry, but you have choices available and it's time to take charge.

How to do it

1. If you notice yourself getting into a worry loop, simply recognize it. When you do recognize it, don't beat yourself up. Acknowledge what you're doing and try to laugh at yourself and lighten everything up. You can even say to yourself, 'Look at me, I'm doing it again!' This is important, because by noticing and light-heartedly commenting on what's happening, you immediately interrupt the autopilot loop.

2. Once you have become aware of the worry, ask yourself, 'Do I want to feel worried/anxious/angry [*or whatever the feeling is that goes with the worry*] or do I want to feel better?' Sometimes, you won't want to feel better right away. I'm sure you can remember a time when a partner or a friend has annoyed you and you felt completely justified in staying in a mood with them, because you were not quite ready to move on from your worry or frustration. If this is you, that's fine, but remember to ask yourself the above question, as it will put you back in control of your process rather than at the mercy of it. Acknowledging that you're choosing to stay in the less positive space is the first step in truly moving beyond it.

3. If and when you make the decision to feel better, release the thought and think about something that makes you smile, such as someone or something you appreciate or that makes you feel a little better. You could imagine lying in a lovely warm bed, or if you're being kept awake during the night, perhaps you could think about how great it feels to be on a beach with the sun on your face. For some of us, thinking about a lovely piece of chocolate or a nice memory or someone we love will do the trick. Once you have found your 'sweet spot', keep reaching for a better feeling or thought, until you can sustain that good feeling.

You don't have to go from feeling terrible to feeling amazing for this to work. Any slight improvement on your mood or distraction from your worry even for a short period is a win. And as with everything, the more you do it, the easier and more instinctive it becomes.

When to do it

Use this technique whenever you notice an anxious or worried thought. The first few times you try this technique your mind will keep bouncing back to the negative place and the whole concept may feel disjointed. However, success comes with continuing to choose to focus your attention on a more positive thought – which works most effectively when the new thoughts are completely unrelated to babies, pregnancy or giving birth. Although this isn't easy, it's worth it.

This exercise is like any other workout: it takes time to get your positivity muscle working harder than the automatic Negative Nelly who is so used to her job that she can do it without thinking. Once you've got it though, what you'll notice is that by regularly practising and reaching for a more positive feeling, you'll instinctively become more positive about your birth and life in general.

To sum up, your new three-step process to move from worry to chill is, in the following order:

1. Acknowledge

2. Release

3. Reach for a thought that makes you feel better

Chapter 5

Creating a Positive Birth Environment

Now that we've covered a lot of the psychological groundwork for creating an empowering attitude towards pregnancy and birth, let's turn to the more practical details.

Choosing where you're going to give birth will hopefully be a fun and interesting process. I encourage you to weigh up the pros and cons of all of your options. This chapter will guide you through the various choices available to you, and will empower you to find the best solution for you.

You might instinctively know that bringing your baby into the world at home feels the most natural option – and then again, you might read that sentence and think, 'You have got to be joking!' The beauty of The Calm Birth School approach is that there is no right or wrong place to meet your baby for the first time. The most important thing is, in the words of the amazing childbirth educator Sarah Buckley, that you feel 'private, safe and unobserved'. When you look at the theory behind that, it all makes perfect sense.

The cocktail of hormones generated when a woman goes into labour are the same as those created during lovemaking. Can you imagine getting intimate with your partner and every five minutes having someone walk into the room and asking you in a loud voice if everything was OK so they could check your progress and give you a mark out of 10 with a clipboard in hand? Getting up close and personal with your vagina while making sure the very non-romantic strip lighting was turned up as high as possible? No? I thought not.

We also have to acknowledge we are mammals and there are certain common traits to any mammalian birth. If an animal is diurnal (awake in the daytime) they tend to go into labour at night, when things are quiet and the daily hustle and bustle is done. Have you ever seen a cat give birth? If so, you've been very lucky, because cats will typically search out the quietest place in the house so that they are unlikely to be disturbed. You might also notice that a number of the birth stories throughout this book are from women whose labour either started or took place during the night.

Our genetic make-up means we are programmed to want exactly the same thing. However, few of the more commonly used birthing environments in the modern Western world offer us a space where we feel a sense of privacy or even quietness. Think of it this way: if you wouldn't want to make love there, your body is not going to vibe when birthing there without taking a few simple but effective actions to make your space your own. I'll get to that in a moment.

First things first: in order to create a positive birth environment wherever you're choosing to birth, I recommend the following four-step process:

Step 1: Know your options

The options available to you will depend on local medical facilities and whether you're prepared to travel to a particular hospital or birthing unit. However, most women can choose from four birth scenarios:

1. Home birth

If you're enjoying a low-risk pregnancy and all the signs indicate that your baby is healthy and happy, you may opt for a home birth. NICE (the National Institute for Clinical Excellence in the UK) has stated that for second-time mums, home births or low-tech midwife-led units provide better birth outcomes in terms of safety and lower rates of intervention than hospital births. In the USA, only 1 per cent of all mothers choose to have a home birth, but that figure is steadily increasing.

In attendance, you'll have a community midwife, who you may or may not have met before; a caseload midwife, who you'll have grown very familiar with during visits throughout your pregnancy; or a private midwife. Being in your own environment, with a familiar care provider, helps to promote the optimal environment for giving birth.

2. Birth centre

Birth centres are run by midwives and aim to create the same type of feeling or environment as being in your own home or, even better, in a spa, in the same way as hospital midwife-led units do. The main difference between a birth centre and a midwife-led unit is that birth centres often stand alone, i.e. they are not connected to a hospital. Once again, these are appropriate for women who are experiencing low-risk pregnancies. If you're based in the UK, there are both private and NHS-run birthing centres available at the time of writing. If you're based in the USA, many birth centres are currently covered by medical insurance.[10]

3. Hospital midwife-led unit

For many low-risk women, this is the middle ground between a birth centre and a labour and delivery ward. This choice offers a place where midwives are familiar with natural births and understand that you're looking for labour to progress at its own pace, with the knowledge that the hospital is close by. This can be the ultimate reassurance for a mother-to-be, helping her to feel at ease should special circumstances arise.

4. Labour and delivery unit

A traditional hospital birth is the number-one option for women who have health issues or are experiencing more complicated pregnancies. If your labour and birth progress without issue, your care will be predominately midwife led. But the labour ward offers consultant-obstetric-led care. Should you request stronger pain relief, such as diamorphine or an epidural then generally speaking, these drugs can and will be administered here.

In the USA, many hospitals offer family-centred care: private rooms where you can go through labour, delivery and recovery all in the same room. After birth, your baby stays in your room with you.[11]

☙ Katie's birth story ☙

My birth story started when I was 40+3 weeks. It was 6:30 p.m. and we were about to have dinner when my waters released. Needless to say, I ended up eating dinner sitting on the loo. The surges were manageable so my husband and I went for a walk and then put on TCBS affirmations and played them on repeat for most of the evening.

At 11 p.m. I called the midwife while my husband inflated the birthing pool. The TENS machine was a waste of time; only the breathing and concentration work helped. I imagined waves every time I breathed through a surge. The in-between time was such a relief.

It felt wonderful to get into the water and from then the time seemed to go quickly – I couldn't believe it when the sun came up! And that's when the surges became really intense. The only description I can give is that they were like the power of thunder; I experienced a really strong juddering that took my breath away.

By 2 p.m. the following day things started to slow down and the midwives suggested I get out of the pool for a bit and have a rest. I dozed on the sofa through surges until eventually they had to use a catheter to help me go to the toilet, so we could remove anything that was preventing baby from moving down my birth path. By that point, I had been in labour for about 12 hours and the midwives were keen to get things going, so I ended up walking up and down the stairs and even doing lunges – which aren't fun at the best of times but especially having done zero exercise for four months! I didn't do the wave visualizations any more; by this point, just breathing through the surges was all I needed.

Eventually, the midwife asked me to actively push because of the time and my increasing heart rate. My husband held me as he sat on our sofa, and I gave birth to our gorgeous little girl, Eliana, at 9:37 a.m. on 18 August. All the way through she had a perfect, consistent heartbeat. She was so calm. The midwives all commented on how well hubby and I worked together as a team and how I managed to literally breathe her out with minimal damage.

I'm so, so grateful to TCBS and everything we learned with you, and how accessible it was to my husband too, who was an amazing birth partner. I never thought a birth like this could be possible, but it really is true! Thank you so much.

Step 2: Ask Questions and investigate

Once you have a sense of where might be the right place to give birth, you can start investigating in more detail. It's vital to give yourself permission to ask questions. If, for example, a hospital labour and delivery ward feels like it might be the safest place for you to give birth but you would still like to birth naturally, I would advise you to:

- Look up the hospital's statistics for natural births vs. unplanned/emergency C-sections.

- Find out whether the hospital supports natural labours and births.

- Investigate what their general policy is for a woman who is on the borderline for gestational diabetes (if relevant).

These are all important questions that, depending on the responses, will either make you feel reassured that this is the right place to give birth, or not.

Step 3: Get to know your environment

If you're going to be birthing in an unfamiliar environment, ask to be shown around the unit. This helps to reassure your subconscious on the day that there is nothing wrong because you have been there before.

If you think about it, before getting pregnant, when would you be most likely to visit the hospital? When you were feeling sick or injured or when someone else was sick or injured. This isn't exactly reassuring for your subconscious. Therefore, if you're choosing to birth in a hospital, getting familiar with the environment and the care providers during a time when you're feeling relaxed and confident signals to your subconscious that there's no need for you to either consciously or subconsciously be on high alert once the time comes for you to visit when you're in labour.

While it's completely normal for your labour to slow down when you move from your familiar home environment to the hospital or any new environment, you can help to lessen the impact if you get familiar with where the action is going to happen beforehand (in order for the subconscious to double-check there is no imminent danger in store).

Step 4: Make it your own

I mentioned earlier that the most important things for you to feel are private, safe and unobserved (*see page 55*). Regardless of where you plan to birth, you want to create an environment in which you and your birth partner feel as private as possible and completely safe. I often have clients include a sign for the outside of their hospital room door stating 'Hypnobirthing couple. Quiet please and please knock before entering'.

While you might be cringing at the very thought of doing this, I urge you to put your concerns to one side and remember that you only get to do this once. So if a sign on the outside of the door is going to remind your care providers to speak in lowered tones and of the type of language you'd like them to use, then it is worth it. These seemingly small shifts are really important in helping you to create a birth experience where you feel calm, private and respected.

Think about your birthing environment as a kind of love or birthing nest. If you're birthing at home, what would be good to have on hand to make your experience as comfortable as possible? If you're birthing outside of the home, what can you take with you to help you to feel relaxed and happy, bringing with it the familiarity of home? This is important because home will be the place where the vast majority of people will feel safest, so recreating that sense of security wherever you're birthing can only serve you.

Here are some ideas for making your birth place feel as comfortable and as homely as possible – make sure you include them in your list, if they appeal to you, and let your birth partner know exactly what you want and need on the day (more on this in Chapter 9):

- Fairy lights – always top of the list – and LED candles

- Fluffy towels for post-birth (your birth facility will have towels, but if your thing is supersoft luxury then this may be for you)

- Pillows

- Room spray/face mist

- Pictures of loved ones

- iPod docking station for music or MP3s (although some hospitals and birth facilities will have these, so check ahead before bringing yours from home)

- Handheld fan

- Massage oils

- Portable blackout blinds (also great for once your baby arrives)

Certain objects will trigger happy and comforting memories for you, so start noticing which items you can consciously and deliberately create positive associations with in your home environment that you can then take with you into the birthing room. Some of my clients create their own birthing playlist, which is a great idea. If you want to do this too, I recommend taking the time to listen to your playlist and practise your breathing techniques while listening before you go into labour. Listening to this out loud means that your baby will enjoy it, too!

Others like to spray their rooms with their favourite scent. Lavender is very calming and soothing, and spraying it around your bedroom in the evening before listening to your MP3s is a great idea. By the time you go into labour, you'll have a really strong association with your chosen scent, which you always smell when you're feeling relaxed and at ease.

Chapter 6

Choosing Your Care

Choosing who is going to support you as you welcome your baby into the world is an important part of your preparations. You want care providers who are going to make you feel special, cared for, understood and respected. Having these core needs met will contribute to how you feel on the day you give birth. The more at ease you feel with the people who are going to be supporting you, the more likely it is you'll be able to create a positive birthing experience.

Professional care

During all stages of care, including the day you give birth, it's important that you like and feel safe with the team you're working with. If you do not feel positive about your midwife, obstetrician or consultant, you're entitled to ask for a different member of staff to work with at any stage.

Some people think it will be too embarrassing to ask for someone else, or don't want to be known as 'that difficult hypnobirthing couple'. They might worry that there won't be enough staff to accommodate their wishes. A phrase that has always served me well when working with my face-to-face clients is, 'What other people think of you is none of your business.' Certainly, when it comes to staffing, it's not your responsibility. The only thing you need to concern yourself with is feeling safe and comfortable, so you can feel positive both during your pregnancy and, crucially, during your birth, where one of your primary tasks is to help your body to work efficiently. If a member of the care team causes you to feel tense, anxious or even angry, this can slow your labour down as your subconscious attempts to protect you and your baby from the perceived threat in your environment by triggering fight-flight-freeze mode, the consequences of which we covered in some detail earlier in the book.

The main point here is that you only get to do this once. So what do you think is more important: putting up with support that is actually hindering the natural process of birth, because you don't feel at ease? Or ensuring that every single person in your space is able to serve you in the way that is going to be most conducive for you to enjoy the birth experience you desire?

Who will be in the room with you?

There's an old wives' tale that you can put an extra hour onto your birth for every person who's in the room. Now while I don't think this is true, feeling private is important. However, rather than focusing on the number of people in the room, we're going to focus on how safe you feel.

What can you do to help yourself feel at ease? Simple: have your dream team surrounding you.

Primary birth partner

Ideally, this is a person who knows you really well. They will have been supporting you throughout your pregnancy and you'll have worked through this book together, understanding the techniques and discussing what will be most helpful on the day.

Managing your birth environment is one of the key roles of the birth partner, which we'll cover in more detail in Chapter 9. In brief, when your birth partner is in tune with you, they can sense anything causing you to feel stressed. The best way for your birth partner to heighten their awareness of how you're feeling is by being present and paying close attention to you.

Part of the birth partner's job is to respond to the subtle, and sometimes not so subtle, changes in your body, noticing whether tension is creeping into your posture or whether you have a clenched jaw or worry lines across your forehead. If a particular person is triggering these things, that person should be asked to adjust their behaviour or leave your environment without question.

Some of my clients have a trigger word or sign to let the birth partner know they need to do something, whether it's a look, a wink or a completely out of context whisper of an agreed password. They know they are needed.

Your birth partner is also your advocate. They are by your side and on your side, and should be the first port of call when managing your environment, so you don't need to worry.

It's important you feel confident that your partner can do their job effectively, as in an ideal world it is best to be disengaged from your rational and logical thinking mind during birth unless absolutely necessary. This is because the minute you feel the need to get involved with stage-managing your environment, the more likely it is for you to be unable to stay in the state of mind that provides you with both the emotional and physical capacity for calmness that you need to birth more quickly and comfortably.

I have been asked before, 'But what happens if I need to be able to make a decision about my birth?' And the answer is then of course you make the decision. Hypnobirthing doesn't mean that you lose the power of control or speech at any time. But for situations that don't fall into the special circumstances bracket, rather than having people asking you questions that constantly put you back into left-brain, problem-solving mode, it would be great for you to be able to trust your primary birth partner to do that job on your behalf.

Doulas

I love doulas! Doulas help to build women's confidence and, as experts in childbirth, they will help you to feel relaxed, more comfortable and safe while barely having to say a word. They offer unconditional and consistent support to you and your birth partner, providing better physical and emotional birth outcomes for both mum and baby. Doulas can also be of great help post-birth, too. Be aware that most doulas specialize in either being a birth doula or a postnatal doula.

A report outlining the benefits of continuous care[12] – this is care from the same care professional throughout your pregnancy and birth – was unequivocal in its findings:

- 31 per cent decrease in the use of Pitocin (synthetic oxytocin used during induction)

- 28 per cent decrease in the risk of C-section

- 12 per cent increase in the likelihood of a spontaneous vaginal birth

- 9 per cent decrease in the use of any medication for pain relief

- 14 per cent decrease in the risk of newborns being admitted to special care

- 34 per cent decrease in the risk of being dissatisfied with the birth experience

Why is having continuous care so powerful? Because it makes you feel safe and secure. Unfortunately, midwives are often unable to give you the type of continuous support that would be ideal as they are working with multiple clients and have huge amounts of paperwork to fill out while you're birthing.

You might be wondering at this stage, 'But what about my partner?' A doula is not there to replace your partner at any stage of your journey, but a doula is likely to have more experience in birth than your partner and has been trained to support you both. A doula will notice the slightest stresses and provide you with the reassurance that you're going to be just fine in a way that a birth partner may not be able to do quite as effectively.

I read an amazing analogy about a doula being like a Sherpa[13] and it really stuck with me. Imagine you're about to undertake climbing a beautiful but challenging mountain with your partner. Of course you want to have them there, experiencing every step of that climb with you, BUT you also want to have your Sherpa, someone who has climbed the mountain many, many times before. A guide who is tuned in to the changing winds, has tricks and tips up their sleeve for getting you past any tricky bits and has the confidence of experience that they are transmitting to you via osmosis. Having your birth partner in the room with you, if you're both keen for that to happen, can be magical and nurturing for you both. However, the statistics are clear: women who are supported by doulas (both with and without birth partners being present) often experience better birth outcomes.

If you can't afford a doula, don't be afraid to approach mentored doulas who are still earning their stripes and charge significantly less. Or approach doulas local to your area and ask if they offer any concessions – you'll be surprised at how many do. Stay open to finding a great alternative who will support you for a significantly lower investment because they are available.

If you know that you don't want a doula, fear not! Millions of women give birth every year without doulas and have fantastic birth experiences. I have given birth twice very positively without a doula, and used one during my last pregnancy – although my birth was super quick so she didn't quite make it while I was labouring, the support in the lead-up was invaluable. So as always, tune in to what feels good and if a doula isn't what you feel you require, then it's all good.

Named or private midwife

Some regions in the UK and the USA offer a named midwife, which means women can build a relationship with the same person throughout their pregnancy and birth. As the report

I mentioned earlier (*see page 68*) showed, continuous care helps you to feel calmer and more at ease during labour.

Imagine the difference it makes to the subconscious, sitting down with the same midwife throughout your pregnancy. Having them get to know you and your family intimately, as well as all of your wishes and concerns. Knowing everything they need to know about you. How reassuring is that going to be? And if you're not birthing in a hospital, they will be the person who helps you to guide your baby out into the world. The peace of mind the right independent midwife can help to provide is – for many women who have used them – priceless.

Mum/mother-in-law/sister/best friend

While some of you may cringe at the thought of your mother-in-law being at your birth, for others it won't be out of the realms of possibility.

Neither is right or wrong. The only thing you need to consider is whether the people you're thinking about are up for the job. Are they able to put their own needs aside, and put you and your beautiful new baby at the centre of their focus, while giving you the kind of support you most want and need? If you feel confident in having a family member or friend as your birth partner, then great. If you're not so sure, perhaps put them on the 'first to know when baby is here' list.

Holding the space

Whoever you choose to support you during your birth, the main job facing them is to 'hold the space' for you. They will best be able to do that if they feel prepared and calm and at least have an understanding of what you have been learning in your preparation. Inviting your birth partner to practise the breathing techniques is not only for your benefit; it will be great for them, too, helping them to stay calm and relaxed during your birth, and ensuring they are not sending out any subtle subconscious vibes to you that there's something to worry about.

Chapter 7

Communicating with Your Care Providers

One of the best things you can do to create a positive pregnancy and birth experience for you and baby is to make sure all the services and advice you receive are specific to you and your situation. This means asking questions of care providers in a way that provides you with very specific information to your unique pregnancy and birth, rather than accepting general rules of thumb that may not be applicable to you. This is important both in the lead-up to birth and on the day.

It can be intimidating when a care provider tells you that you need to take a specific course of action, which you either don't understand or feel is unnecessary. Having the confidence to ask *why* and *how* certain things are relevant to your specific situation can make the difference between creating a positive experience and feeling like birth was something that 'happened to you'. I will share a specific set of questions you can use later in this chapter.

When you practise asking questions during pregnancy – such as 'What makes you think that about me in particular?' or 'What would best suit me in this situation?' – you'll gain clarity, which will allow you to make informed decisions about how you would like to progress. Having a sense of control is a key barometer for wellbeing. This need is particularly heightened during pregnancy and birth.

While sometimes you may conclude the best thing to do is relinquish control and move forwards with your caregiver's advice or opinion, when that decision is made by you (armed with facts that are relevant and specific to your pregnancy), you remove doubt and confusion and know that the decisions you made enabled you to get the right birth for you and your baby on the day.

Use your BRAIN

In order to access an experience of feeling calm, empowered and in charge of your birth, I invite you to lead and direct the conversation using the acronym BRAIN to get clear on what is being requested or suggested by your care provider. You can then make your decisions in an informed way and as a team.

B – Benefits

What are the benefits to me and my baby of moving forwards with the suggested action?

R – Risks

What are the risks to me and my baby of moving forwards in this way? What are the risks of not moving forwards?

A – Alternatives

What are my other options?

I – Instincts

What is my instinct on this, my gut feeling?

N – Nothing

What is the immediate risk if we choose to do nothing for the next 30–60 minutes? (Be specific about the time period.)

When it comes to your pregnancy and birth, clarity is king! So remember to use your BRAIN and give yourself the opportunity to make informed choices.

> ∾ Tip ∾
>
> It is also a good idea to include in your birth preferences that unless there is an immediate medical need for either you or your baby, you and your birth partner would like to be given the time and space needed to discuss your options.

Birth preferences

You'll communicate with your care providers via your birth preferences, commonly referred to as your 'birth plan'. I deliberately don't use that term when talking about your birth choices, as plans can often feel more definitive – and from the stories you've already read so far in the book, you'll know that nothing is definitive when it comes to birth! Making plans for a wedding totally makes sense, as it would be completely inappropriate not to arrange all the details in advance rather than express preferences: 'If you don't have enough salmon for the wedding breakfast, don't worry, we'll have beef.' However, birth can't be neatly packaged up in the same way as a wedding can.

Be wary of veering too far in the opposite direction, however. While there are no guarantees in birth, the idea of not completing a birth plan at all because nothing ever goes to plan (a comment that I hear frequently) completely misses the point. After talking to the doulas over at Doula UK, I would recommend that when you discuss your birth preferences with your birth partner, you take an A, B, C approach.

Think about what you would like to happen during a straightforward birth (Plan A), how you would like things to proceed if you need some assistance (Plan B), and how you would like to be supported should you end up in a situation where an unplanned C-section is the best course of action (Plan C).

Taking this approach means that while it is important to spend your time and energy focusing on your ideal birth scenario, you have also taken into consideration the practical elements that may arise should you not have a straightforward delivery. It serves you, your birth partner and your baby far more powerfully from an emotional perspective if you can navigate the 'unexpected' from a place of preparation, rather than having to make knee-jerk decisions on the spot. And while you can never plan for every eventuality, understanding the importance of using your BRAIN will also help you immensely to turn difficult situations into infinitely more positive ones.

~ Tip ~

For ideas on what to include in your birth preferences, please use the following link to download our birth preferences planner: www.thecalmbirthschool.com/bookbonuses

Ana's story below is an example of how even when birth unfolds in an unpredictable way, you can choose to feel calm, empowered and in charge.

✎ Ana's birth story ✎

My babies' journey into the big wide world started when I was taken into a delivery room with two lovely midwives who had read my birth preferences and said they would do all they could to help me achieve them.

My waters released on Friday morning at 4:25 a.m. and the waves began pretty quickly. They grew in intensity over the next four hours. I used a TENS machine and my calm breathing and wave breathing, which worked brilliantly to help me stay in control and enjoy the experience, as I knew each one took me a step closer to meeting my babies. Unfortunately, I didn't dilate so I was offered a hormone drip, which I politely declined, and we were able to go another four hours in the belief that the babies just needed a little more time. Then, when I still hadn't dilated, I agreed to the hormone drip.

The surges started to come thick and fast, and I continued using the TENS machine and breathing techniques along with a mat, a bean bag and a ball, which all helped me to relax, flop and go into my birthing zone. I have to mention here that there was no offer of 'pain relief', as I

had made this clear in my birth preferences, and I greatly appreciated it because I felt that my care providers had faith in my ability to birth.

I used very deep noises as the surges peaked, which I didn't expect to do, but it helped massively – as well as counting how many breaths I needed to get me through each wave. This was a saving grace because I knew roughly how long each wave would last (although some were a lot longer or shorter).

My husband had watched The Calm Birth School videos with me so he was an absolutely brilliant birth partner, giving me drinks, space or comfort when I needed them. When the waves got really intense, the best 'pain relief' was holding on to him. It must have been the oxytocin! Anyway, after 16 hours I still hadn't dilated and I had a temperature so the doctor (who was brilliant at stopping talking to me when another wave appeared!) suggested increasing the hormone drip.

This was such a hard decision for me because it wasn't something I wanted. Along with saying yes, I knew that I would need an epidural (not once was this suggested to me. I felt in total control). However, I knew, using the BRAIN process, that the benefits outweighed the risks: the fact that I'd been birthing for so long with little effect meant that if I did nothing, it could be an extremely long process, which would tire us all out.

I asked for an epidural to help numb the intensity of the surges from the increased hormone drip. This is where it gets interesting and I believe it was all meant to be. Firstly, because of the TENS machine and using TCBS breathing techniques, I didn't feel the epidural go in; secondly, it meant I could have a break from the intensity.

Most significantly, an hour after having the epidural, I was told that twin two's heart rate was worrying and that it was beneficial to deliver the babies as soon as possible – again the care providers left it to me to deduce that they meant a C-section, allowing me to feel that I was the one in control, not them.

So after 20 hours of labour during which I'd been able to use all of my well-practised techniques, I was taken to theatre where my beautiful babies were born. And do you know what? It was such an amazingly calm, relaxed, light-hearted atmosphere. Everyone was so lovely, informative and respectful. Again, because I knew that my birth could take any twist and turn, I had so many tools in my toolbox from The Calm Birth School, and my husband and I were able to embrace whatever avenue our babies' births took. I can honestly say it was one of the best days of my life. Yes, it was intense, challenging (I had to dig very deep) and long. Yes, unfortunately I wasn't able to have everything on my birth preferences (delayed cord clamping with twins hadn't been done there before and as it was classed as an emergency C-section, we agreed that it was best just to get the babies out).

However, I felt empowered, strong, confident, positive and calm throughout. The breathing techniques, positive affirmations and confidence skills I had developed from The Calm Birth School had no doubt allowed me to enjoy the birth of our babies and despite not having them naturally, they were born into a happy, calm atmosphere, which is what we wanted. I felt like I changed that day, not only by becoming a mother, but also because it was the first time I had truly committed to something that no one around me had done before, where I really believed in something and I succeeded. I'm incredibly proud of myself and can use the skill set I now have to take me into motherhood. Thank you, TCBS, for all your positivity and support: we couldn't have done it without you.

Chapter 8

Preparing for Labour and Birth

We've taken a look at the emotional side of birth, but we all know it's a mammoth physical task too, so in addition to practising your breathing techniques daily, how else can you help your body and mind to prepare for the big day? This chapter will walk you through some key areas to focus on as you prepare for your labour.

Pelvic floor muscles

The female body is magnificent, perfectly designed to create and hold a baby for 40 or so weeks. However, I'm not going to try to kid anyone – growing a human being puts a lot of stress and strain on our bodies, even when we come from a natural default setting of being fit, strong and healthy. One of the areas that quite literally takes its fair share of the load is the pelvic floor.

Your pelvic floor is made up of the muscles that run from the base of your pubic bone to the back of your spine. Imagine them shaped a little like a hammock or a sling keeping everything else in place. Most of us won't think about stress incontinence in relation to ourselves – even for a moment – until pregnancy kicks in, when people suddenly can't stop going on about how important the pelvic floor is. Looking after your pelvic floor during pregnancy means that you:

- Are less likely to leak urine after birth.

- Are less likely to experience vaginal prolapse (when your pelvic organs begin to bulge into the vagina).

- Will boost your post-baby sex life.

> ~ Tip ~
>
> If you would like an in-depth look at what you can do to increase and protect your pelvic floor, check out our Pilates Masterclass with Dr Joanna Helcke, at www.thecalmbirthschool.com/bookbonuses. However, a great starting point for your pelvic floor preparations are the exercises outlined below.

Pelvic floor techniques

While you may be familiar with the old school squeeze-and-hold exercises, things have moved on considerably since then. Claire Mockridge, a Nutritious Movement™ certified Restorative Exercise Specialist (www.clairemockridge.com) explains below what's changed and offers the most effective way to prepare your pelvic floor for birth.

Up until a few years ago, pregnant and postnatal women were advised to perform traditional pelvic floor exercises known as 'Kegels' to help strengthen the pelvic floor ready for birth. These exercises were invented in the 1940s and involve 'squeezing and releasing' the muscles up to 30 times a day, and 'drawing up and holding', working up to a 10-second hold. However, Kegel exercises are more suitable if you spend most of your day on your feet, whereas many women today have a tight pelvic floor musculature from years of excessive sitting, and doing so in very poor alignment.

Basic anatomy tells us that the pelvic floor muscles run from the pubic bone at the front of your pelvis, to the tailbone at the back. And if you sit with your tailbone tucked underneath you, you're effectively moving the tailbone closer to the pubic bone and passively shortening your pelvic floor. So learning how to sit correctly for optimum pelvic floor health is a must.

Being in a seated position for work, rest and play takes its toll on the body: your hamstrings, hip flexors and calf muscles all become tighter, and because you're sitting on your gluteal muscles (butt) these aren't getting a workout at all (and neither is your pelvic floor).

Ideally, you'll want all the muscles that feed in and out of the pelvis at their ideal length and strength for the pelvic floor to function well and enable everything to 'give' during childbirth.

Gaining flexibility into the hip flexors and hamstrings with regular stretching exercises and also strengthening the gluteal muscles by performing plenty of squats, lunges and other butt-building, body-conditioning exercises is really a good starting point.

Is it possible to create length in the pelvic floor to enable you to give birth with as little damage as possible? Yes, it is, and the following two exercises will create length in your hamstrings, help you keep your pelvis mobile and ease those aches and pains in your lower back and pelvis.

∼ Hamstring stretch ∼

The hamstrings attach into the back of the pelvis, so lengthening them helps move the tailbone away from the pubic bone, which in turn opens up space for your baby to move through the birth canal.

How to do it

1. Standing, set your feet hip-width apart with your toes pointing forwards.

2. Place your hands on a table or on the back of a chair and move your pelvic weight back behind you. Imagine you're reaching your sit bones away from you/relaxing your hipbones down towards the floor.

3. Drop your rib cage down towards your pelvis and stay here for a few minutes. Keep your arms straight and your shoulders relaxed.

When to do it

Do as many reps of this exercise as feels comfortable or for about 10–15 minutes three to five times a week.

∼ Pelvic rocking ∼

Another great movement to strengthen your pelvic floor is one that you can do anywhere when standing, (even in labour) to help keep your pelvis mobile, release tension in your inner thighs and lengthen the pelvic floor muscles.

How to do it

1. Place your feet wide apart with the outside edges of your feet straight.

2. Position your hands on a chair, table or kitchen countertop and, as you tip forwards at the hips, reach your sit bones away from you.

3. Keeping your legs straight, gently rock your pelvis over to the right, then the left, feeling a lengthening sensation in the inner thighs.

4. Drop your rib cage down towards your pelvis.

When to do it

Do as many reps of this exercise as feels comfortable or for about 10–15 minutes three to five times a week.

Perineal massage

The perineum is the area between the opening of your vagina and your anus. When you're labouring it gets extremely thin as it stretches, so try massaging your perineum to help prevent any damage and reduce the likelihood of tearing or episiotomy – especially for first-time mums.

⟿ Perineal massage technique ⟿

This doesn't tend to be a favourite with mums-to-be and I can't understand why! All joking aside, the more familiar you are with your vagina, the more connected you'll feel to yourself and your baby. So don't feel embarrassed. If you find manoeuvring around your curves and bumps awkward then ask your partner to get involved or wait until you're in the bath. However, if you feel totally uncomfortable with the idea of perineal massage, you might like to try an excellent product called Epi-no. You can find out more information about it here: www.epi-no.co.uk

How to do it

Start by washing your hands and then lie down on your side, adding a few drops of a vegetable-based oil, such as grapeseed or almond oil, to liberally coat your thumb or index and middle finger.

Insert your fingers into your vagina and gently massage and stretch the area in a U-shape motion.

If you have difficulty reaching the perineum from this angle, standing up and placing a foot on a chair in front of you is a position that many women find provides them with easier access.

When to do it

Start perineal massage from around 36 weeks into your pregnancy and do it for about five minutes once or twice a week.

Prenatal bonding

Prenatal bonding is a hugely important part of the process of becoming a parent. From a mother's perspective, it helps you to tune in to your baby and your body, which is invaluable when you're birthing because it helps you to feel even more of an instant connection with your baby. For partners, it is an excellent way to create a bond with their baby before birth, which not only helps to create a sense of involvement for the partner during pregnancy, but also helps them feel more of an immediate bond with baby once it has arrived.

It's not unusual for some partners, particularly dads, to take up to six months to bond with their new babies. The time you spend preparing as a couple for your baby's arrival can be hugely beneficial for the attachment process, and prenatal bonding can help to take things to the next level.

Prenatal bonding involves acknowledging and interacting with your baby on a playful and emotional level while they are in the womb. It's easy to recognize our babies are growing physically within us, as our stomachs look fuller every day, but tuning in to how our babies are growing on an emotional level is not something we always consider.

While we can start prenatal bonding at any time, the last trimester is when our baby's brain development kicks into overdrive. This is the time when our babies are laying down all of their instinctive patterns of behaviour as they spend 80 per cent of their time in the REM (rapid eye movement) state. REM is the state of consciousness you enter whenever you're learning something new. Child psychologists talk about what happens between birth and three years of age being crucial, as our children will spend 60 per cent of their time in the REM state, downloading much of what is going on around them, creating the templates they will live their lives by. By the time we are adults we only spend around 20 per cent of our time forming these templates, unless we utilize tools like hypnosis, which allow us to enter this state of conscious, helping us to learn new patterns of behaviour more quickly and easily.

During the last 12 weeks of pregnancy, your baby will learn to look at a human face as opposed to an animal's face when they are born. They will learn how to instinctively grab hold of your finger if you place it in their hand, and they will also know they can mimic and copy, which is why if a person pokes their tongue out at a newborn within the first 24 hours, many babies will return the favour. Being able to mimic, engage and hold on are all vital skills for survival, as they help to promote bonding and attachment with parents immediately post-birth.

Babies communicate with us through their movements, responding to our thoughts, emotions and our external environments, which are all inextricably linked, as far as I'm concerned. Perhaps this is why mums often report that a baby who has been relatively quiet throughout the day will start kicking and stretching like crazy whenever they start listening to their MP3s. I like to think this is baby giving the nod to all the wonderful endorphins coursing around the body.

This is also why some mums worry about the impact of stress, anxiety and anger on their babies. While it is true being consistently exposed to stressful situations during your pregnancy is not ideal, it's not a good idea to beat yourself up every time you experience a cross word with your partner or find yourself in a stressful situation. In fact, a little bit of a difficult episode is the perfect time to practise your TCBS techniques. Being able to instinctively turn to your breathing exercises that help you return to a feeling of emotional calm when faced with those unavoidable difficult or challenging situations takes practice. So embrace these situations, as it's your ability to tune in to that calming space within that is exactly what you're going to need to do when you're birthing.

You are also teaching your baby the same valuable skill.

While we all instinctively want to protect our children from any difficult emotions or encounters, real life means they will face these situations regularly. It's likely that at some point during your pregnancy you'll feel stressed or frustrated with your partner. This generates stress hormones. However, when you call upon your Calm Birth School tools, your baby gets to experience a return to calm as your body responds with endorphins when you choose to focus on your breath. The lesson you teach your baby here, as they already start to develop their own emotional intelligence, is that even after a storm things always return to a state of calm. The message is that everything will always be all right, which is a very beautiful gift.

So what can you do to help with this bonding and learning now?

Relax

Every time you access a state of relaxation, listening to your audios and visualizing your baby being born, you're strengthening the bond and connection between the two of you as well as creating future memories for your hard drive.

Talk and read to your baby

Baby can start hearing your heartbeat and distinguishing between sounds and voices from around 23 weeks. Getting your partner to talk to or read to baby regularly is a great way for them to start getting familiar with each other.

Massage your bump

Another great way to get your partner involved is gently massaging your bump while talking to you both. Your baby gets to associate your partner with the feel-good hormones, endorphins.

Play games

Responding to your baby's kicks can be both reassuring and a great way to have fun with your baby before they arrive.

Sing

They don't care if you sound like Madonna or not: they find dulcet tones reassuring, so sing it out sister!

Music

Playing the same songs so baby becomes familiar is another great helper once baby arrives, in them being settled: they will associate your favourite pregnancy songs with being safe and secure in the womb.

Understanding estimated due dates

As you start thinking specifically about the big day, it's worth turning your attention to your guess date, or as the professionals like to call it, your estimated due date (EDD). I totally get it – being pregnant for 40 weeks, particularly if you're one of the early birds who knew as soon as you conceived, is a really long time! This happened to me with my son, Caesar. But eventually, the day arrives: your magic estimated due date (EDD).

If you're like most of the pregnant population, your ankles have somehow merged with your calves, your rings no longer fit your fingers, you haven't been sleeping well for weeks or months, you're tired of wanting to go to the lavatory every hour, only to realize you can barely squeeze out a wee that would fill a pipette, and turning over in the middle of the night feels like trying to shift the Taj Mahal. I remember it well. Your EDD shines down on you like a beacon of hope. Not only is it the day you supposedly get to meet your gorgeous little bundle of joy, but also the many, many other benefits that come with giving birth make you cling on to your due date like a limpet on a rock face.

The problem with this, though, is that 96 per cent of us are being led up the river without a paddle. For most of us, the day arrives and... nothing: nada, zip, diddly squat. This is for a number of good reasons. The 40-week EDD is based upon Naegele's Rule, a theory developed in 1744 by Hermanni Boerhaave, a botanist. Boerhaave came up with a method of calculating the EDD based upon evidence in the Bible indicating that human gestation lasts approximately 10 lunar months. The formula was publicized around 1812 by German obstetrician Franz Naegele and since then has become the accepted norm for calculating the due date. However, there are more than a couple of massive question marks in Naegele's theory.

Strictly speaking, a lunar (or synodic – from new Moon to new Moon) month is actually 29.53 days, which makes 10 lunar months roughly 295 days, a full 15 days longer than the 280-day gestation we've been led to believe is average. In fact, if left alone, 50–80 per cent of mothers will gestate beyond 40 weeks. I've always gone against conventional wisdom and insisted that pregnancy is a 10-month process.

While your period may have managed to sync with all the girls in the office, the reality is that we all have different cycles. Even when you know the date of conception, there is a five-week variance with healthy mothers giving birth to healthy babies. So if you give birth at 37 weeks, you're not three weeks early, and if you give birth at 42 weeks, you're not two weeks late. You are only postdate once you exceed 42 weeks.

You might be surprised to learn that only 4 per cent of women give birth on their due dates, which is why I prefer to call it a guess date. Remember the power of language (*see pages 20–21*)? From the moment you're told your guess date, I strongly advise you do your best to forget it. If you find that the midwives at your antenatal appointments are slightly obsessed with it, don't worry: it's their job. Make your peace with knowing it, but perhaps don't share the specifics with your friends and family. The last thing you want is people texting and asking, 'Are you still pregnant?' which is beyond annoying. Instead, start telling people that you're due around the middle or end of the month (or even two weeks after your guess date, if you want to say a date!), to relieve yourself of any additional pressure. This might feel like you're being dishonest, but it's really an act of self-care, so that as you approach and possibly pass your guess date, you're not being hounded with messages and phone calls checking up on you.

∽ Laura H's birth story ∽

The day before my guess date (Saturday), I woke up around 1 a.m. thinking I might be having mild cramps, but they were so mild I wasn't sure if I was imagining it or if it was general pregnancy aches. I felt so excited that I couldn't get back to sleep, so around 4 a.m. I got out of bed and watched a movie. I went to the toilet a few hours later and saw what I thought was my mucous plug. (I realized later that I had only lost a very small part of it and there was a lot more to come!) We texted our midwife to let her know that there was some action but that it would probably be a while yet, as the cramps were still very mild.

I spent the day watching movies, walking, bouncing on a fit ball, listening to MP3s, using the TENS machine and even visited some family. By the evening the surges were stronger and closer together, so I was pleasantly surprised how quickly and easily it was all happening. We were timing the surges and by 2 a.m. they were around three minutes apart and quite strong, but not unbearable. We called the birthing centre and although the midwife was sceptical that I was ready, she agreed we should come in.

By the time we had driven the 20 minutes to the birthing centre my surges had all but stopped and I was a bit embarrassed that I'd overreacted. The midwife said I was in very early labour and offered me an internal exam. She was surprised to find I was 2cm dilated and completely effaced, so I was further along than she had thought. She said my waters were bulging and to expect a gush any minute. She sent us home and said she'd probably see us back there in a couple of hours.

At home, I got in the bath while my husband slept and had some intense surges. This was probably the lowest point for me – although the surges were still not unbearable, I was starting to wonder how much stronger they would get and how long it was going to take.

By morning my surges had slowed right down to 20 minutes apart. I called the midwife and she told me to try to sleep, but I couldn't as every time I lay down my body would cramp, although I was fairly comfortable bouncing on the fit ball, walking or in the bath. I was starting to feel tired, I couldn't hold down any food or water and it felt like I was going backwards. The rest of the day was much the same with very mild surges spaced far apart.

Finally, at 10 p.m., the surges started to intensify and quicken again, and by 1 a.m. we were back at the birthing centre where I was very pleased to hear that my midwife was available. I had some anti-nausea medication and I started trying to rehydrate (I hadn't kept down any food or water for over 24 hours at this point).

I got into the birth pool and spent the next few hours breathing though my surges, which were quite bearable. My husband was really great, chatting between surges, feeding me ice and water, and counting the breaths for me. I intended to use the gas and air, but kept waiting until the surges were unbearable, which they never were. At one point the midwife mentioned that we'd been there three hours and I honestly thought it had only been one! I was progressing quite slowly but was feeling relaxed and the baby's heart rate was stable.

By 6:30 a.m. (Monday) my waters finally broke and I felt the need to start pushing. I pushed a few times and it felt like the baby was about to pop out, then I looked down mid-push and realized I wasn't nearly that close. It took another hour of pushing before the head was out as my surges were about five minutes apart. This meant that I got lots of rest in between, but the baby was retreating after each surge.

When the head was finally out, the midwife asked if I could push the baby out before the next surge, so I stood up and pushed while the midwife helped guide our baby out. As soon as she handed me our beautiful baby girl, I held her straight to my chest and within 15 minutes she was latched on and feeding.

Bonnie is now three months old, feeds like a champion and is a total mumma's girl. After the birth I felt so empowered and had such faith in myself that I think it really helped with breastfeeding and caring for my baby. Although I don't remember it, the midwife later told me that straight after the birth I said, 'I could do that again!

∽

Chapter 9

Your Birth Partner's Role

When I used to teach hypnobirthing to couples face to face, I would always joke that the birth partner was supposed to be a bit like Batman: with a utility belt full of tools, techniques and knowledge about how to support you best on the day you meet your baby. Many of the women who take The Calm Birth School video course love the fact that their partners can watch the videos and get a full picture of what they can do by learning exactly what they, as mothers, are learning.

The role of the birth partner begins long before labour. If you haven't been sharing the tools and the philosophy of the book, make sure that you start sharing it now. Your birth partner's energy on the day you go into labour will have a profound impact on how you're feeling, even if this is at a subconscious level.

While the majority of birth partners will be men, this information also counts for same-sex couples, mothers, sisters, friends too – anyone who is going to be your primary support.

It's quite easy when we look at the stereotypes of birth to feel that it is a one-woman show. And while women are the ones giving birth, how we feel in the lead-up to our baby's birthday and how we feel on the day is very much a team effort. So to reiterate, make sure your partner reads this chapter if nothing else. This is especially for them.

During pregnancy

The gift of listening is one of the biggest gifts your birth partner can give you. Being heard during your pregnancy and birth is one of the biggest dictators of how a woman perceives her experience to be. And while it's very easy at times to think your partner knows, or should already know, what you want, often they don't. This is not their fault; it is just one of those things. So getting in some dedicated listening time is a priority. The byproduct of this will be confidence that your birth partner knows what you want and will be capable of working with your care providers to help you stay calm and positive throughout. And perhaps most importantly, it will foster a deeper connection between the two of you.

If you already feel you have a great foundation for listening, use the next exercise to build on what you have already, and if you know that you and your birth partner could use a little help with getting on the same page when it comes to your wants and desires for your birth, then this is the perfect opportunity to create the space in which to connect.

∼ Listening with intention ∼

The great news about this technique, as with all the techniques in this book, is that it is not rocket science but it will take a little effort from both of you to make it happen.

How to do it

Take some time out together. We've talked about the importance of having some dedicated time for yourself every day (*see page 14*), but spending time with your partner talking about the birth is magical. And while the emphasis in the earlier weeks is likely to be on birth choices, how you're feeling and any worries or concerns, this is also a fantastic forum for both of you to discuss your expectations, plans, ideas and any worries or concerns you may have about being a parent for the first time or again.

How to do it

The only rules are 1) no technology, and 2) when one of you is talking, the other person must listen.

Here are my top tips for listening with intention:

- **Don't interrupt:** Don't look for the next opportunity to say what you think, even if you agree with what your birth partner is saying – be present.

- **Make eye contact:** Reflect what your partner is saying to you before you add your personal comments to make sure you have heard and understood what has been said.

When to do it

Aim to sit down together at least once a week, and if your birth partner is away during your pregnancy, then downtime together can be just as effectively achieved over Skype or FaceTime. If you're unable to have any contact with your birth partner during your pregnancy, do not stress out about it. Just make sure you prime them once they return so that they can support you by reading this chapter of the book. It's also a good idea to write a list of any questions, concerns or things that are important to you or that you would like them to consider, so you can discuss them once you get the opportunity.

Learn the breathing techniques

I can't emphasize enough how useful it is for your birth partner to learn the breathing techniques with you (*see page 203*). Your birth partner will know you better than anyone in the room, after all the quality down time you have been spending together, and is the person with whom you're likely to have the strongest emotional bond.

There is no doubt you'll pick up on the energy they are emitting while you're labouring and birthing. If they are anxious, worried or concerned about you or the process, they will be producing their own adrenaline and cortisol, which they will then have to focus on managing the fight-flight–freeze response (*see page 8*) rather than on you and your needs. They may also be unable to respond or manage what is going on outside of themselves as effectively.

The biggest problem with your birth partner feeling stressed is the potential for you to pick up on those emotions. It's a bit like walking into a room in which a couple have been arguing, where you can feel the tension and negative energy lingering even though nothing is being said. If you notice that your partner is feeling stressed, either consciously or subconsciously, the impact can cause you to lose focus or, on a subconscious level, fear there is something to worry about. So having a birth partner who can manage their own range of emotions is paramount. And, as you already know, one of the best tools for helping you to regain a sense of peace, calm and control is to regulate the breath.

If your partner is instinctively taking long, slow deep breaths because of all the practice they have done and radiating a sense of calm and peace, you're far more likely to pick up on it and adjust your breathing accordingly, without them needing to tell you to do anything differently. This allows them to guide you into relaxing and lengthening your breath without any need for a conversation.

It can also help to avoid unwittingly pushing your buttons by saying things like 'remember to breathe', or 'are you doing CBS breathing now?' if they just take the lead and set the example to you by demonstrating long, audible deep breaths.

Massage

We'll look at the power of positive touch in a later chapter because, although you may not want this type of touch during labour, if you become accustomed to being massaged by your partner during pregnancy, the positive associations will help you in labour – even if all you do is hold hands. If you get used to visualizing birthing your baby while your partner is touching you during pregnancy, then you can strengthen the association between you and your partner when it comes to your baby's birth. At least 10 minutes a day of light touch massage is just what TCBS orders.

Bag-packing

Write out a checklist of all the items you think you might need during your labour. This is a really positive thing to do (even if you're birthing at home) for the following reasons:

- You'll have everything you need all in one place.

- If you change your mind and decide you would prefer to birth in hospital or special circumstances arise that make sense for you to move locations, your birth partner won't miss anything vital off the list.

> ∼ *Tip* ∼
>
> Get your birth partner to pack the bag for you. This isn't being lazy but will help your birth partner to know exactly what's in the bag and where to find anything you might need. It's also useful to keep the checklist on the top of the bag, so should you ask for anything, your partner can quickly scan the list to check that it's in there, without needing to ask you.

On the day

Throughout labour and delivery, remember that your birth partner is your servant. No joke. Your birth partner is responsible for ensuring that you're super comfortable in early labour and that you have everything you need.

Early labour

The early phase of labour – particularly if your birth partner will be coparenting with you – can be a fantastic time for bonding and connecting before your baby arrives. Use these moments to tune in to each other – and the fact that life will never be quite the same again – in the best possible way. This part of your labour can be deeply, deeply romantic and intimate, especially first time around.

Your excitement should be palpable and you can, if you want, turn this part of your pregnancy and birth into a beautiful moment. You might choose to go for a walk together, light some candles, listen to music, enjoy a massage, look into each other's eyes and celebrate what's happening. If that sounds too soppy for you – don't worry, find your own way to take stock and connect. However, this bonding is important, because the more you can feel as though you, your baby and your partner are working together, the more oxytocin you can't help but produce, which often means an easier labour. Your birth partner can play a vital role in helping you to generate oxytocin. This may be by making you laugh, offering reassuring pep talks, kissing, hugging, massage, slow dancing, nipple stimulation or any of the above that will raise your mood. If you decide you would prefer to take this time for yourself, of course that is OK, too. Just focus on spending time on activities that make you smile, and increase your endorphin and oxytocin levels.

So what can you do with your partner to make this moment a moment?

If this is baby number two or more, it can definitely be more challenging to create the type of intimacy I have outlined above, but don't be defeated. If your older child or children are present, consider what you can do to make this a family moment instead. Perhaps it's still going for a walk in the park, letting them know that their baby brother or sister will be here sooner rather than later. Or maybe ask a friend or family member to babysit, so you don't need to focus on keeping them entertained. There is no right or wrong answer to what to do with older children during labour and birth. Just do what makes sense for you and the family.

While not quite as romantic, other practical things your partner can do are to remind you to use your birth ball, make sure your musical playlist is ready and that you have funny films on tap. They should have all of your liquids and snacks ready, and know where the birth bag is too, of course.

> ~ Tip ~
>
> Be mindful of not going into deep relaxation mode straight away in early labour, it's totally fine for you to keep active for as long as feels comfortable for you. Just make sure you have enough snacks, drinks, your birth playlist and MP3s on hand, and that you are comfortable.

At the hospital

Once you're at the hospital, your birth partner can make sure your care providers have a copy of your birth preferences and have read through them. Your birth partner should know where the spare copy has been tucked away, too.

While you settle in, your birth partner can get busy making sure that the birth suite is as comfortable as possible and that everyone in the room is in tune with your wishes. Should they notice someone who isn't on the same page as you, they should ensure they speak to the consultant midwife or whoever is in charge to see if they can allocate another member of staff to your team.

Other practical considerations might include:

- Lowering the lighting or putting up LED candles around the room.

- Placing pictures in optimal positions.

- Setting up your MP3 player so you can listen to your music or TCBS audios.

- Breathing with you slowly and audibly.

For the most part, your birth partner will play the role of the silent observer unless you're up for having a chat in between your surges, of course. They will also:

- Be on the lookout for any build up of tension in your face or body and assisting you by remembering to breathe themselves.

- Be ready to hold you or touch you, as hand-holding or light touch massage can be great to help support you during a surge, if you're open to it (*see also pages 149–150*).

- Whisper words of encouragement.

- Encourage you to drink fluids regularly and eat to keep your strength up.

- Remind you to go to the lavatory every 45 minutes or so, as babies do not like passing full bladders.

- And, most importantly, they remember to be present.

ཉ **A dad's birth story** ཉ

Like many men, I knew very little about pregnancy, birthing and babies, and had never heard of hypnobirthing before attending classes. To say I was sceptical about the concept is an understatement, but I quickly changed my ideas after attending a taster session. I'm not entirely sure what it was that changed, but I'm confident that hypnobirthing gave me the tools necessary to support Rachel, my wife, in having a calm, positive birthing experience.

For me, the journey started before the classes. Finding out that we were expecting our first child was hugely exciting, so much so that I didn't really consider what I needed to know and do in preparation for the arrival of our baby. I didn't realize the extent to which Rachel was anxious about giving birth – something that I now know many women face as a result of years of media focus on the negative aspects. This could've eaten away and hindered the birthing experience that Rachel desired from the outset. It was during the first hypnobirthing class that the penny dropped and I became aware of my responsibility to be more actively involved. This wasn't Rachel's pregnancy; this was our pregnancy.

Hypnobirthing equipped me with a greater level of knowledge about birthing in general. No question was ever a stupid one and I found that if I was unsure about something, no matter how trivial, the other dads in the group were likely to be unsure as well. This helped me to better support Rachel in the latter stages when we were faced with some special circumstances. I could stay calm and work with Rachel and our medical team to make the right decisions for us and our baby. We didn't always get the decision that we wanted, but we knew that if our desires were challenged, it would be for the good of our baby.

Hypnobirthing enabled me to have a positive impact throughout our pregnancy but, most importantly, 30 minutes before Sebastian was born I could give Rachel some additional support with her breathing and pushing. Faced with some discomfort, Rachel started to panic and the consultant was keen to perform an instrumental delivery, something that we wanted to avoid. I knew from our classes that if needed, additional support might be required but that breathing techniques could and should be used first. I could calmly communicate with the consultant while supporting Rachel with her breathing and pushing techniques, which resulted in Seb being born without any intervention. Without these techniques, Seb would've been born in very different circumstances.

Finally, hypnobirthing armed Rachel and me with a level of empowerment that would have been impossible without attending classes. Most notably, in putting together our birthing preferences, we made them personal to us by including photographs of us and our pets so that the midwives and consultants would have a better understanding of who we were. This doubled up as a great comfort to Rachel when she started to miss her home comforts. All the midwives involved in the delivery of Seb commented on our birthing preferences document and this filled us both with confidence that our wishes would be considered where possible.

While these aspects stand out as very helpful for me, my greatest insight was the understanding that it was OK for unforeseen circumstances to challenge our birth preferences. Hypnobirthing was never about doing sufficient preparation to guarantee a predefined set of outcomes; it was more about using the techniques to overcome challenges and still achieve a positive outcome in calm surroundings. I found that even though we weren't able to follow most of our birthing preferences, hypnobirthing gave us the knowledge and unity needed to remain calm, even though special circumstances arose.

Chapter 10

Induction and Special Circumstances

Remember, the thing that gets your labour going is oxytocin – the hormone of 'luurve' (*see pages 38–39*). If your guess date has been and gone and you're getting tense and frustrated because your baby doesn't seem ready to make an appearance, remember your estimated due date is a guess date (*see pages 90–91*). Then have a word with yourself and partake in some oxytocin-inducing, relaxation-friendly activities.

However, sometimes for reasons and situations beyond our control, special circumstances arise before our guess dates or we require assistance from our care providers during labour. It is in these special circumstances that all of your training in staying calm, using your BRAIN and really leaning into the idea that you can still enjoy a positive birth experience, even if you're not getting your Plan A preferences, come into play. And when it comes to hearing the many different birth stories, it's often these stories that have the biggest impact on us as women describe how they were able to navigate the twists and turns of labour from a position of strength and control.

Continuous monitoring

Depending on the nature of your pregnancy, you might find yourself in a position in which you're invited to have continuous monitoring before or during your birth. This situation may occur if:

- You have high blood pressure.

- You have or develop pregnancy-related diabetes.

- Your waters have released and more than 24 hours have passed.

- You have had a C-section in the past.

- Your waters have released and there is meconium (baby poo) in the water and the baby doesn't appear to be stressed.

- You are expecting twins.

It is incredibly important that you use the BRAIN system (*see pages 74–75*), and ask your care providers about the risks and benefits of being continuously monitored in your specific situation, and particularly if the reasons are non-medical – such as having had a previous C-section or if you are expecting twins, for example. You do have options. The latest research shows that while continuous foetal monitoring reduces the chances of neonatal seizure in utero, it also increases the chances of having a C-section or assisted delivery.[14] So please do your homework. You always have a choice.

Pre-labour interventions

The reason that so many people spend time discussing the arbitrary 40-week EDD is because this date plays a big role in determining whether or not a woman will be advised to have or offered a pre-labour intervention, and if you're hoping for an intervention-free birth this is a big deal. However, when it comes to pre-labour interventions, I want you to put the 'shoulds' out with the rubbish.

If there is no medical reason for you to need intervention, i.e. you're a low-risk mother experiencing a low-risk pregnancy, but you reach 41+3 weeks and decide you just have to meet this baby and would like some assistance from your medical team to help this process along, then this is your right to choose to do so, just as it is your right to refuse it.

Sometimes what you want will be moral support to remind you that you really are on the home straight and will be meeting your baby soon enough. At other times you'll just want to get on with it. And it's during those times that you may be presented with the following options:

Stretch and sweep

Depending on the local policy of your care providers, once a woman passes her EDD she may be offered a stretch and sweep. This is where your care provider will use their fingers to massage the neck of the cervix, in the hope of stimulating the uterus and kick-starting labour.

As with all interventions, a stretch and sweep can only be carried out with informed consent so you'll need to make up your mind as to whether you feel happy with moving forwards with a sweep or not. Depending on where you're based, this intervention can be offered prior to 40 weeks. Some women decide that they do not want any form of intervention at all; others feel that when faced with a possible chemical induction they would rather give the 'less invasive' option a try. Neither decision is right or wrong. This is your body, your birth and your baby. You have to do what feels right for you.

However, although being offered a sweep is common, research shows there is no significant statistical evidence to support membrane sweeps inducing labour.[15]

Induction

If you're a low-risk mother experiencing low-risk pregnancy, it is unusual to be offered an induction prior to your EDD. However, an increase in blood pressure, a prolonged start to labour after waters breaking, or concern over baby's growth rate, are all reasons that it may be recommended.

While hospital policy will vary from region to region, once a woman reaches 41+1 weeks, and particularly if she is birthing in a hospital, her caregiver is likely to start discussing the option of booking in an induction.

Inductions are an extremely emotive subject within the birth world. I interviewed Virginia Howes, author of *The Baby's Coming* and independent midwife, who has supported thousands of couples during pregnancy and birth. Virginia firmly believes that unless there are specific medical reasons, women should absolutely not be induced. You can listen to Virginia's interview by visiting www.thecalmbirthschool. com/bookbonuses

My personal view on induction is not so black and white, and this is simply because I don't think being petrified of induction is helpful. I have received many letters from Calm Birth School students who have used TCBM and experienced incredibly positive births after induction, because they were able to stay calm and in control throughout.

Having said that, choosing to be induced is a big decision, as a medically managed birth means that your body doesn't produce its own oxytocin and endorphins in the same way as a birth that is able to progress naturally. The consequence of this is that many of the benefits we talk about in terms of being able to enjoy a comfortable and more relaxed birth are more difficult to experience. Often the process makes birth much more intense. However, anecdotally, hypnobirthing mothers who have trained themselves to relax deeply on demand and can go with the sensations they experience as opposed to fighting against their body, are in the best position to create a more positive induction experience. Go back to being clear on the risks and benefits of moving forwards with any procedure during pregnancy and birth, and make your decisions based on what feels right for you and your family.

If you do decide to move forwards with an induction, my key piece of advice is to build up your bank of endorphins and get your natural oxytocin level as high as possible before going to the hospital. Put the following three priorities at the top of your induction checklist:

1. **Feel happy.** Once you know you're going to be induced, you also know that within a couple of days you're going to be holding your baby in your arms for the very first time. Woohoo!

2. **Listen to the Fear Release MP3:** Acknowledge any fears or anxiety about being induced that you might be holding on to and then let them go.

3. **Make more endorphins:** Start building up your bank of endorphins by spending time with your partner, and doing activities that make you feel happy and relaxed, whether that's making love or cuddling your partner, watching funny movies or drawing.

> ∾ Tip ∾
>
> Check out the suggestions for creating more oxytocin in Chapter 9 (see page 101), if you're in need of inspiration.

Building up your bank of natural endorphins so you feel more relaxed and at ease before your induction will help you manage the stronger surges more effectively so you can work with your body as opposed to against it. And should you start surging regularly, you can always request that you continue to labour without further stimulation. (I recommend putting this in your birth preferences.)

And to reiterate: you'll still be able to use the techniques outlined in this book should you go in for an induction. In fact, it's very important that you do and because you'll know your date for induction, you'll have a head start! So use this to your advantage.

The induction process

Most women will have been offered up to two cervical sweeps prior to a medically managed labour. Often a gel or a pessary containing artificial prostaglandins to help to ripen the cervix and stimulate uterine waves is inserted into the vagina. If this is unsuccessful, you'll be asked if you're open to having your waters broken. This is usually done with a small hook that looks similar to a crochet hook. The impact of artificially rupturing your membrane can, in some cases, mean your labour progresses very quickly and intensely. If this isn't enough to get things moving, you'll be offered syntocin or picotin, depending on where you are in the world, which is a synthetic form of oxytocin, administered via intravenous drip.

Many women find the pressure waves stimulated by these artificial drugs are much more intense than the body's own natural waves. When induced, the body does not respond by producing endorphins in the same way as when a woman goes into labour naturally, which is why your own preparation, should you be having an induction, is so crucial. If you do have a medically managed birth you'll be offered an epidural to eliminate the pain. This is something some women choose to move forwards with immediately while others choose to wait it out. There is no right or wrong answer to this; only you can decide what is appropriate for you and your family on the day.

If you decide induction is not for you, this too is a valid choice, as you do not have to move forwards with any medical care, including being induced. This can often feel like a difficult decision to make, which can leave you feeling very isolated, unless you're under the care of an independent or private midwife, because you'll be going against the status quo and your care providers may be against waiting for nature to take its course.

Often, women are told they run the risk of their bodies not functioning as effectively if they allow their birth to continue past the 42-week period, alongside risking the health of their babies.

This is something no mother-to-be wants to hear and many tend to proceed with the intervention despite feeling uncomfortable, as if they do not have alternative options. However, the risks when stated in this fashion do not present a full picture, which means mums-to-be are unable to make an informed decision. I strongly recommend that if you're thinking about induction and are not sure whether it is right for you, you visit www.aims.org.uk (Associations for Improvements in Maternity Services). This charity provides robust statistical, evidence-based research that is extremely accessible for those of us without medical backgrounds. This will allow you to assess whether induction is the right decision for you or not.

Hypnobirthing is about understanding what you can control and letting go of what you can't. In terms of medical interventions, you're always the person in charge. No one can prevent you or 'not allow' you to do anything – and you always have the right to make a decision based on informed consent. You have the right to choose not to move forwards with an induction. You have the right to decline vaginal examinations. You can choose to birth at home after a previous caesarean. You have the right to ask for evidence to support the advice you're given, and there are many sources of information that you can seek out independently to help you to make an informed choice about how you would like to proceed with your pregnancy and your birth.

> ## ∼ Tip ∼
>
> When faced with any non-urgent medical matters, use your BRAIN system — Benefits, Risks, Alternatives, Instincts, Nothing (see pages 74–75). Ask questions and do the research so you can look back and know you had the best birth experience for you on the day.

If you decide an induction is not for you, but you would like to feel even more at ease with your decision, you can request additional monitoring of both you and your baby. You can also ask for the amount of fluid around your baby to be checked, as well as baby's positioning.

Bear in mind that we were not designed to stay pregnant. And barring special circumstances, our body gets it spot on. So if you're a low-risk mother enjoying a healthy pregnancy and all the signs point to a healthy baby, many people would argue your baby hasn't been born yet because your baby is simply not ready. Your body will kick into action when your baby is ready.

Special circumstances

We have already discussed birth preferences and their importance, particularly in situations when birth does not go to plan (*see Chapter 7, pages 75–76*). While it is extremely important to focus on creating the calm, positive birth you desire, sometimes even with the best planning in the world, special circumstances may mean you'll have to look at your B and C options for birth preferences.

Sometimes women who end up having pain relief for the intensity of their surges or perhaps meet an unplanned scenario feel like there has been some kind of failure on their or their body's part. I want to emphasize that no one can fail at birth and your aim is to enjoy an experience that you look back on with joy in your heart, whether your baby arrives vaginally or breech, with the help of forceps, an epidural or via caesarean.

All of these experiences can be as magical as the next when you feel clear on why certain precautions or actions are taking place, and when it feels right for the health of you and your baby. The aim is always for a positive birth, not a perfect one.

Alongside the possibility of birth not playing out exactly as you anticipate, for what can be a multitude of non-medical reasons, it's also important for you to be mindful of the clear medical reasons why you should seek support during your pregnancy or labour should you notice any of the following:

- Feeling your surges start prior to 37 weeks. Although it's very common to experience Braxton Hicks – a tightening of your tummy without much intensity in the latter stages of pregnancy – if you're surging and noticing intensity before the 37-week period, you should call your care team straight away.

- If you notice a change in your baby's regular patterns of movement after 24 weeks.[16] Many people mistakenly believe that baby's movement will slow down towards the end of their pregnancy, but this isn't the case. Take a look at www.kickscount.org.uk for more information on how to monitor your baby's movements.

- A prolonged period of time from your waters releasing and the onset of labour. I include this tentatively and urge you to do your own research around this, as many women are offered antibiotics or inductions on the basis of safety for both mother and child when, in fact, the evidence to support the normal course of action for most medical care providers suggests that waiting to see if labour progresses normally without any intervention poses significantly less risk than what is often presented.[17]

- Noticing that your membranes are discoloured or have a strong odour once they have released. This could be meconium (baby's poo), which can be a sign of distress for baby, while an odour might indicate a urinary tract infection.

- Regular and severe headaches.

- A very high temperature or fever.

- Extreme vomiting or diarrhoea.

- Blurred vision or dizziness, particularly if combined with substantial swelling of your hands, face, ankles or feet.

- Heavy bleeding. While it's normal to experience spotting towards the end of term, your care team should immediately check out any heavy bleeding.

As always, don't forget to tune in to and trust your instinct; if something doesn't feel right, even if you can't put your finger on it, receiving reassurance from your care team that everything looks OK and there's nothing to be concerned about is a great endorphin producer, which will always be a good thing for both you and your baby.

Helping birth along naturally

When your baby is almost ready to make an appearance, sometimes it is possible to give them a little encouragement and help them on their way. While I don't believe any of these suggestions will help if your baby's not ready to make an appearance, they can be useful if all baby needs is a little nudge in the right direction. Or if you feel like you 'need' to do something, anything, to encourage this baby out of its incubation period... of course, you don't need to do anything at all because nature's got you covered, right? But trust me – I know at 11, 13 and 15 days beyond my guess date with each of my bambinos – I know that feeling of just 'wanting to get the baby out!' So here are a few things for you to consider trying to encourage them.

Sex

Sex and/or physical intimacy is arguably one of the best things you can do to nudge things along. When we get intimate, particularly when we are climaxing, we produce huge amounts of oxytocin. So touching, kissing and making love are all fantastic ways to encourage labour to start. Prostaglandins, also found in men's semen, help to soften and ripen the cervix, which in theory can help with dilation.

Reflexology

Reflexology is great at around 36/37 weeks, as a reflexologist skilled in maternity reflexology will stimulate certain pressure points within the body that are known to help start labour. Rather than waiting until the final minute, as a last resort think about a course of sessions around the 36-week mark, which anecdotally are reported to be more effective in helping women to meet their babies closer to their guess date.

Acupuncture

A series of sessions from around 36/37 weeks can be very helpful. Some practitioners offer specific packages for avoiding induction, too. Acupuncture originates from Chinese medicine and the methods used work with the body's energy systems and meridians, moving a woman's energy to help baby make an appearance.

Acupressure

This works in a similar way to acupuncture but without the needles, and focuses on stimulating labour by applying pressure to points across the body.

Being physically active

Walking up stairs and steep hills is great, because you tend to be in a forward-leaning position, which helps to get your baby to put more pressure on the cervix in order to help stimulate the uterus into action.

Dates

Eating six dates a day from 36 weeks onwards has been shown to increase the likelihood of spontaneous labour and reduce the chances of intervention.[18]

Nipple stimulation

Using either your hands or a breast pump will help to produce the hormone oxytocin, which you'll definitely know by this point is great for stimulating birth.

Spicy food

Any spicy food that might make you want to go to toilet, basically. Kick it up a notch. So if you're used to eating a madras curry you need to go for a vindaloo or a phal, whereas if you're normally a korma girl then a madras would be fine.

(... Oh, and did I mention sex?!)

∞ **Emma's birth story** ∞

I was sent for a growth scan at 36 weeks and baby was measuring fine, but the amniotic fluid was on the low side, so we had to go back for monitoring every couple of days. Everything was fine, but then I was told that since I was 37 weeks, why wait? I was originally supposed to be having a home birth, so this would have been a big change for me and I wanted to wait. We asked for another scan and daily monitoring.

The second scan showed that the fluid level had reduced even more, so we agreed a date for induction that worked for us in terms of arranging childcare for my first child and coordinating when our doula would be available. I ended up being induced at 38+5 weeks. I did lots of research about what would and could happen, researching pain relief and telling the midwives that I would ask if I needed something. The midwives were very positive about trying to make it as natural as possible. We had packed some fairy lights and my affirmation posters and decorated the room.

I had two pessaries. The first did nothing and my cervix was not nearly ready, so we walked for miles around the hospital grounds. The second pessary was put in at 3:30 p.m. and I started to get mild cramps about an hour later. I was examined at 6 p.m. and my cervix had just started to dilate. Things started to get a bit more intense, so I put the Birth Rehearsal MP3 on repeat and tuned in to my breathing.

My doula arrived just after 9 p.m. and the midwife examined me and said I was 4cm dilated, which to be honest still felt like I had so far to go. I refocused and every time I had a surge, I just breathed and imagined waves crashing against the sand.

At some point I asked for gas and air, and a bit later I felt I needed an epidural, mainly because I was really tired and there wasn't a break. The anaesthetist arrived after what felt like forever, but in hindsight I was in transition. It was about midnight at this point. I was finding it very hard to listen to what he was saying and feeling an intense pressure. The midwife insisted on examining me while I was having several surges and said I was 8cm. Another surge I was 9cm and then one more and I was 10cm.

Baby Ophelia Isabel was born at 1:24 a.m. with minimal tearing and no epidural. I was vaguely aware of the anaesthetist leaving the room at some point!

From start to finish, the whole process was 7.5 hours, much more intense than my first 36-hour directed, purple-pushing, forceps labour, but so satisfying to know that I could deliver completely naturally. I was much more in the hypnobirthing zone. I think, in a way, because I was induced I knew it was coming, and I started my breathing straight away rather than being too far gone like last time. During my first labour, my doula had had to work hard to talk me through each surge. This time, I was chatting then I would have a surge and then pick up conversation again. It was amazing, and the scripts and general calmness of The Calm Birth School really helped.

Chapter 11

Caesarean Birth

One of the biggest misconceptions I have wanted to bust through my work at The Calm Birth School and using TCBM techniques is the idea that women who end up having a caesarean birth – either unplanned or elective – have in some way been let down by their bodies. And in the case of unplanned caesareans they have been let down by hypnobirthing, or if opting for an elective caesarean they don't think they will benefit from hypnobirthing. I'm here to say that all of that is rubbish.

Unplanned caesarean

Historically, unplanned caesareans are often referred to as emergencies. However, 'emergency' is not always the most accurate way to describe events – by the nature of the emotions generally attached the word 'emergency' is not the most useful way to describe the procedure, which is why I like to opt for 'unplanned' instead.

The great thing about unplanned caesareans when you're using The Calm Birth Method is that you *can* plan for them beforehand.

Again, while it is not useful to focus lots of attention on plans B and C, doing so now, getting clear on what you want and what you don't want, will serve you well should anything unexpected happen.

So why might your care providers suggest that a caesarean may be the best course of action for you and your baby?

- Prolonged first or second stage of labour.

- If baby appears to be in any distress.

- If baby has manoeuvred into a non-optimal position that will make it difficult for you to labour optimally.

- If there is any excessive bleeding.

With most unplanned caesareans, the decision to move forwards isn't an urgent one and you'll have time to talk and think it through. This is the time to use your BRAIN (*see pages 74–75*). If the situation is urgent your care team will want and need you to act quickly. This can lead to there being lots of people in the room and lots of hustle and bustle. This is where your breathing techniques really come into play. While you may not be able to control what is going on around you, if you have been practising your breathing techniques in real-life situations, you'll be able to tap in to narrowing your focus of attention, so that you keep your breathing slow and long, feeding your uterus and baby with oxygen.

Elective caesarean

If you're opting to have a caesarean for whatever reason, again the work that you do during your pregnancy will provide you and your partner with a fantastic foundation for:

- Releasing any fears you may have about going into hospital or having the operation.

- Helping you and your partner to take time out to actively and consciously bond with baby and the new roles you'll be taking on.

- Providing you with breathing techniques to help both of you stay calm and relaxed during the operation, to bring your pregnancy to fruition.

- The breathing and visualization techniques you practise during your pregnancy will also help you to manage any pain you may experience during your recovery, too.

Natural caesareans

It's very useful to outline how you would like to experience your caesarean birth in your birth preferences plan. Things you might want to consider:

- If having an elective caesarean, ask to see what the operating theatre looks like, so you know where you'll be having baby.

- Ask if you can play your own music.

- Request if you can wear your own clothing

- Request that your hands be kept free from any monitors.

- Any anaesthetic used should not affect the upper part of the body so you can hold your baby as soon as they have been delivered.

- Any intravenous drips are placed in your non-dominant arm.

- Have a think about if and when you would like the screen lowered. Many women like to have it up for the incision and then taken down so they can see their baby being lifted out.

- Where do you want your birth partner to be standing – close to your upper body or closer to baby?

- Do you want immediate skin-to-skin contact or would you prefer for baby to cleaned up first?

- How long would you like the umbilical cord to be left before cutting? Although you'll have mentioned this in your ideal birth scenario, it is important to highlight it in case of a C-section too, if you would like it to go white or at least stop pulsating naturally.

- Would you or your partner like to cut the umbilical cord?

Having this incredibly gentle start to life, where you dictate how you would like things to happen, can be equally as powerful and soul affirming for mothers as a vaginal birth, but you have to think about what you want and ensure that your birth partner or you can communicate those wishes effectively. Mothers who have caesareans give birth too, and we want you to own your delivery.

Chapter 12

Signs of Early Labour

Early labour is known as the 'latent phase'. Contrary to popular belief, not all of us will 'just know' when we have gone into labour. Some women will, but if you find yourself second-guessing all the usual aches and twinges as you approach or pass your guess date, do not fear. You are not alone. To help determine if this really is your moment, here are seven helpful hints of what to look out for.

The seven signs of early labour

1. Excessive nesting

Most women want to welcome their baby into a lovely home that feels fresh and clean, which is why you might get sucked into all of those drab tasks you've been putting off for weeks. However, if you notice yourself verging on obsessive, this could be a positive sign things are due to start happening. Think of it as you instinctively preparing your nest.

2. Cramping

A universal description for the start of labour is a period-like, cramping feeling, often felt in the abdominals. If this is your only sign, although it may mean things are beginning to move along, it's understandable that you may be uncertain about whether or not it's really time. Doubt often arises if the sensations are irregular, with a complete lack of intensity and no other accompanying signs. As you'll see from Laura K's story in Chapter 15 (*see pages 153–5*), sometimes early labour can take a while.

Some women experience weak cramps for days or even weeks before they move into established labour, while for others things progress more quickly. If you experience cramping, look for consistency in the frequency of the sensations to indicate if early labour is starting. If you're destined to experience a long lead-up to labour, infrequent cramping sensations can be frustrating and confusing. This can be referred to as a long latent phase or practice labour.

A good way to deal with the uncertainty is to practise the acceptance you'll be using while birthing: don't wish for things to hurry up or think about when you're going to know for sure. If you can, go about your everyday business, staying as active as possible. If you can, get outside and go for a walk. Bear in mind when you're on the move that you might want to stay within an hour of your home, just in case. Continue to reassure yourself with the knowledge that your body is getting prepared. Sooner or later you'll be holding your baby in your arms, so relax and go with the flow.

3. Spotting

A little bit of light spotting – often partnered with the dislodging of the mucous plug (but not always) – is fine, normal and a sign your body is preparing for labour and birth. If you notice any blood prior to 36 weeks or feel it is too heavy, it is always worth checking in with your medical care providers.

4. Waters releasing

This is one of the first signs that spring to mind for most people when we talk about indications for the onset of labour. Once again, a reminder about language: rather than talking about waters 'breaking', I encourage you to describe them as 'releasing'. Nothing is broken during labour. It's all good.

What you might be surprised to learn is that the release of your waters is not a given. You won't necessarily experience a big gush of liquid like you've probably seen in the movies. Sometimes this does happen, and it's not unusual for some women to choose to sleep on towels, or have some type of protective covering for the mattress, just in case. Some women will gush for hours on end. If this is you, there is nothing to worry about; you're just releasing a lot of water.

However, many women find themselves unsure as to whether or not their waters have released at all, as they only feel dampness. I call this trickling. If you're a trickler, you're more likely to find yourself wondering whether or not you've wet yourself rather than if your waters have released. Sometimes you'll hear an audible pop when the waters release. Again, this is nothing to worry about.

In some cases, your care providers will want you to travel to your place for giving birth to confirm whether your waters have released or not. Many will ask you to put on a sanitary towel, so they can check for moisture and confirm that the waters are clear and baby hasn't passed any meconium. Meconium is the equivalent of baby doing a poo in the amniotic sac and will turn your clear waters a shade of brown/green-brown. It can sometimes – but certainly not always – be a sign that baby is in distress. Depending on your caregiver and the colour of the meconium – a general rule of thumb is the lighter the colour, the happier your care providers will be – you'll be asked to participate in continuous foetal monitoring, so your team can keep a close eye on baby's stress levels.

Other care providers are happy for you to stay at home and wait for you to travel until your surges are established. Speak to whomever you're working with in advance and find out what the protocol is, so you can make appropriate plans for the big day.

5. Losing your mucous plug

Your mucous plug sits in the neck of your womb and as your cervix begins to get wider, this mucous membrane is discharged. Losing your mucous plug is a fantastic sign that your body is preparing for birth. However, be aware that this can happen up to two, or sometimes even three, weeks before you go into labour.

Your mucous plug can be clear and quite snotty-looking, but it can also be a bit pink in colour and sometimes bloody. The loss of the mucous plug can sometimes be referred to as a 'bloody show', because the mucous becomes bloodstained. This is totally normal and nothing to worry about.

6. Diarrhoea

We love a loose stool at The Calm Birth School. A bit of diarrhoea before or while you're experiencing early cramping sensations is a big green light that your body is expelling everything it needs to, making the path of descent as easy as possible for your baby. Bring on the poo!

7. Feeling nauseous or being sick

Nausea is another fantastic sign your body is doing what it needs to do and is clearing your digestive system, so your body is in the optimal state for birthing.

Bonus sign: Intuition

You might have female intuition going on and sense you need to stay close to home on the day your labour begins. For example, it's not uncommon to hear of women saying even though they had no signs at all, they just knew that they should stay close to home that day.

How to deal with a long latent phase

The latent phase of labour is something that every woman experiences. It most commonly lasts 1–24 hours. While no latent phase is typical because everyone experiences it differently, a good indication you're experiencing the latent phase is getting non-regular surges or cramping. Surges can be very intense for some women, but for others they are so mild that they don't even feel labour has begun.

For women who do feel intensity with their surges, this can be a challenging time, as you just want to know when things are going to kick in properly, particularly if your irregular surges last longer than 24 hours. Yes. You read that correctly the latent phase can last longer than 24 hours – we have had women looking for moral support after working with their body's surges for five days in the past! But of course this is where the Calm Birth Method and your mindset come into play. My top tips for not going insane during this period are as follows:

- Stay active for as long as feels humanly comfortable.

- Get down on all fours.

- Walk sideways up the stairs.

- Check out the website www.spinningbabies.com for more handy hints on positions you can take that may help to move baby into a more optimal position for birth.

- Look for support in the community group.

- Go for a massage or reflexology session to help you relax.

- Appreciate that, whatever happens, you're going to be holding your baby in your arms for the first time soon enough.

- Listen to the MP3s that most resonate with how you're feeling or where you would like your focus to be at the time.

- Remember to drink lots of fluids.

- If you're feeling tired, it's totally OK to sleep.

- You know we love a fabulous warm bath, too.

Expect to run the full range of emotions, from happy and excited in the beginning to frustrated, angry and confused the longer it goes on. All of these emotions are totally normal, so acknowledge them and look to release them as much as possible while focusing on the fact that your body is preparing for your baby to meet you for the first time.

The main thing is to listen to your body.

What to do when labour starts

During the night

If your labour starts at night, rest, relax and conserve as much of your precious energy as possible. You don't want to feel tired just as your body and baby need you to feel energized. In the same vein, if your birth partner is with you and you go into labour at night, let your partner sleep if the sensations you're experiencing aren't very intense and you don't need support at this time. You'll want your partner to be as alert and supportive at the exact time you need them to be and if they're tired because they have been up for most of the time, that's going to be a problem. While there isn't any doubt that you've got the biggest job on the day, being a birth partner is a tiring job – even if all they appear to do is just sit and hold your hand.

Being in the early stages of labour when you would normally be asleep is the only time I would ever recommend you lying down. However, you can still do this with a slightly elevated back, by propping yourself up with cushions, or lie down on your side to create a more optimal position for baby.

During the day

If you're in early labour during the day, stay active but, more importantly, always listen to your body and rest when you need to, in order to ensure you conserve your energy for when you need it most. When you feel a surge during early labour, stop moving and focus on your breath before continuing with your activities. Keep a 'business as usual' mindset. Stay active and keep your body as upright and forwards as possible. This aids the natural birthing process, working with gravity, helping to encourage your baby to stay in, or move into, the optimal position for birth. Look back to Chapter 9 (*see page 101*) for inspiration on how you and your birth partner can keep the vibe and those loving feelings oh, so high.

When it comes to chilling out, sitting down, bouncing on your birth ball and lying on your left side can all be great resting positions. One of your birth partner's roles will be to gently remind you to switch positions every 45 minutes or so. If they notice you have been static for too long, reminding you to move will ensure you do everything you can to encourage baby to move down your birth path optimally.

If any part of you feels apprehensive instead of excited – and this counts during the daytime too – acknowledge the fears, practise your Calm Birth School breathing and release and let go of anything that is causing you anxiety. This will create the best foundation for a calm and comfortable birth.

✍ Natasha's birth story ✍

I want to say a BIG thank you to The Calm Birth School and everyone in the Facebook group who offered advice and support during my pregnancy. Thanks to TCBS I was able to have the birth experience I had imagined. My son, Buddy Jude Middleditch, was born on 10 September and I'm so in love with him! The breathing techniques, MP3s and TCBS wisdom helped me to feel confident and prepared for anything.

I was overdue and went into labour the day before I was due to be induced. I was able to stay at home and work through my surges before going to our local birth centre. I found the affirmations incredibly helpful and had them on repeat. I wrote some of the positive affirmations down on cards and added some things that I knew would keep me going, so my husband could read them to me. This really helped; hearing his voice made me feel safe and got me to relax. I think this helped my husband to come up with things to say when I was struggling.

When I was in the birth pool I remember him telling me to think of the baby's tiny hands and feet, and to imagine the fun we would all have together. The other thing that helped me during each surge was to imagine waves coming in and out. During the final stages I imagined I was a lion on the shore and to push the waves back out I'd roar. This may sound strange, but it really helped me to maintain my breathing and focus.

TCBS also helped me to cope with an infection I had after the birth. I'm so grateful for all the invaluable advice and tools that TCBS gave me, and could write pages about it. Thank you so much!

Positions for Labour

When it comes to finding the best positions for birth, I urge you to follow your own natural birthing instincts and go with what feels comfortable. There are no rules as such, but there are certainly positions that can be more or less helpful for creating positive birth experiences.

Active birthing: Sitting, squatting or on your hands and knees

Active birthing was made popular in the eighties through an antenatal educator and birth pioneer called Janet Balaskas. After completing lots of her own research into birth in different cultures, it became increasingly clear that far from being strapped down to beds and often anaesthetized as was common in the West, in other cultures women often kept moving while labouring until they felt ready and would then proceed to squat during the birthing phase. A study in 2012 finally provided supporting evidence that women who remained active and upright during birth had a significant decrease in interventions.[19]

Active birthing allows the pelvis to open, providing your baby with more space to travel through the birth path. There is also evidence that these type of positions lead to shorter labours.

Keeping mobile during active labour

The following positions will help you to stay mobile and as relaxed as possible, in as optimal a position as possible, during labour.

- Sit down on the lavatory with your torso facing the cistern, so you can rest your arms on the back of the toilet. This position can offer great relief, alongside helping you to open your pelvis.

- Sitting astride a chair, with your torso facing the backrest, can have the same effect as sitting on the loo.

- Use your partner to hold on to them around their neck, while gently swaying or circling your hips.

- Sit on a birthing stool, a chair or a ball.

If possible, avoid lying on your back while you're labouring, as it will never help you birth your baby more quickly or comfortably. If you find yourself wanting to lie down on your back, discuss with your partner how they can help you to stay mindful of moving around – without them sounding like they are ordering you about. Give yourself a limit of 30 minutes and then move into one of your favourite upright and forward positions.

~ Tip ~

Check out Katy Appleton from Apple Yoga's bonus video on prenatal yoga and different types of positions at www.thecalmbirthschool.com/bookbonuses

When movement is more restricted

You may find yourself lying on your back if you need to have continuous monitoring or an epidural.

Continuous foetal monitoring can mean that you have to lie on your bed, so the machines can take as accurate a reading as possible. So before you get to this stage remember to use your BRAIN (*see pages 74–75*) to identify if there are any other options available to you, like intermittent monitoring, so you can still walk around, helping you to support your baby's descent more effectively.

When it comes to an epidural you may be wondering what the big deal is, as you won't be able to feel anything. The negative impact of an epidural is less about comfort and more about the potential of you needing further assistance, as again your pelvis will not be in the most optimal position for guiding baby out into the world.

An alternative is to ask if your care providers offer the option to have a low-dose epidural, which will still allow some feeling of sensations and may allow you to adopt a position on all fours, for example. Alternatively, it is possible to ask that your epidural dosage be turned up and down, so you have more sensations when it is time to breathe your baby down. Both options can help to reduce the chances of you requiring a ventouse or a forceps delivery.

❧ Irene's birth story ❧

Damiano was born at 1:28 a.m. on 21 April and he has brought us immense joy and happiness, despite a 23-hour labour!

My labour started on Monday 20 April at 2:45 a.m. when I was woken up with light cramping sensations. At first I didn't think they were surges and I kept going to the toilet and going back to bed. I was 38+5 weeks and thought I had another couple of weeks ahead. I couldn't sleep any longer and at 4 a.m. I woke my husband to tell him I was in labour. He started timing the surges and although they were coming quite regularly, they weren't strong at this point. We kept timing them until around 8 a.m. when we decided to call the midwife, since by this time the surges were regular – every four to five minutes – and lasting for a minute each time.

The midwife visited us at home. I was only 1cm dilated, so she told us to keep doing what we were doing and to wait until the surges were stronger before calling her back. In the following hours I took a nice warm bath, used the ball in all possible positions, listened to music – in particular Adrift *by Christopher Lloyd, which accompanied us till the delivery of the baby – and kept breathing, breathing, breathing.*

At 3 p.m. we called the midwife again as the surges were stronger and quite close to each other. She told us to go to the hospital, as I should have been at the right stage by then. We called a taxi, grabbed everything we needed and went to the hospital.

Once at the hospital, I was really upset when the midwife told me I was only 2cm dilated and needed to go home. I tried to persuade her to let me stay and eventually she agreed we could stay another hour, after which she would assess me again.

I think my body knew I didn't want to go home, because all of a sudden there was a gush as my waters released. The surges were getting stronger and when I was assessed again, I was 6cm dilated.

When we were settled into the room, I had a bath to relax and then spent time on the birthing ball and on the floor, with my husband breathing with me and my sister giving me a nice back massage. The midwife even helped with aromatherapy!

I got in the pool and kept feeling the need to push, but the surges were not strong enough to push baby out. After 40 minutes in the pool the midwife noted that his heartbeat was slowing down, so I had to come out. It was a shame, as the warmth of the pool was lovely.

Once I was upright again, the surges became nice and strong again – so strong, in fact, that I only just managed to make it back to my room and the bed in time. And three pushes later my lovely baby was out!

After a few minutes, the cord was clamped and I cut the cord, which gave me a strong happy sensation. Damiano was placed in my arms and I delivered the placenta naturally within 10 minutes, so the lovely journey of motherhood started skin to skin with my baby!

During all those hours of labour I had no pain relief, no gas and air, nothing. It was just me, my husband, my sister and the two midwives, Charlotte and Suzanne.

Chapter 14

Active or Established Labour

Depending on where you are in the world, active labour is considered to begin from 5cm dilation onwards. Sometimes women report they feel more intensity once established labour begins, some say as soon as they felt their first surge it was all systems go, and others report no difference between the early stages of labour or after they have surpassed the 5cm mark. Remember: everybody experiences birth differently.

If you feel immense intensity straight away, rather than the gentle build-up many expect, remember you have all the tools you need to access a deep state of relaxation quickly. By simply using your Calm Birth School breathing (in for four and out for seven), followed by your wave breathing (in through the nose for seven and out through the mouth for seven), you won't need half an hour to enter a deep state of relaxation. You'll be able to access this state quickly and easily.

However, for this to come instinctively to you, you need to practise breathing techniques regularly in situations non-conducive to relaxation. Don't wait until it's late at night and you're all warm and cosy in bed to practise. Go for it when you're in the middle of chaos, being bumped about on public transport or wanting to explode at your birth partner because they have said the wrong thing... again. When this happens, your body will be used to responding with relaxation when you step into the less familiar territory of giving birth.

When to travel to your care provider

If you're birthing away from home, the closer you are to established labour of 5cm dilated or more, the less likely it is you'll need any medical intervention. For first-time mothers, I recommend that unless you're going to feel safer or more comfortable travelling to your birthing place earlier, stay at home until you experience four waves in 10 minutes, each lasting for about a minute, consistently for two hours. For those of you who have done this before, three waves in 10 minutes, lasting between 45 seconds to a minute, is a good rule of thumb.

> ～ Tip ～
>
> If your birth partner is with you, let them time and measure the frequency of your waves.

Some women prefer not to use a clock to time their waves for a valid reason: whenever we engage with something that requires us to use our rational mind, it is more difficult to distance ourselves from the sensations we are experiencing in our bodies. Having your birth partner in control of the clock can help if you're going down this route. Having some kind of signal or trigger word so they know when to start and stop the timer can also minimize chatting if you're the type of person who prefers to focus all of their energy inwards when they are birthing. Maybe you'll be extremely engaged and talkative in between surges, but until you get to the day, you simply don't know which camp you'll fall into, so it's useful to have a plan.

An alternative method for measuring how far dilated you are comes from the book *The Art of Midwifery* by Hilary Marland, and utilizes the fact the body is diverting more blood away from the legs and feet towards the uterus in order to increase its efficiency while we are birthing. When a woman is 1–2cm dilated, the feet and ankles are colder; once she reaches 4–5cm, the calves are colder; and then at full dilation of 10cm, the legs are cold from the knee downwards. Clever, right? Be aware that this is not a technique you can use if you have been getting in and out of the shower or bath, since the warm water will change your body temperature.

～ Tip ～

If you're birthing at home, talk to your care provider about when would be the best time during the process for you to contact them before they come to you.

Whenever you experience a surge, use your Calm Birth Method breathing techniques to eliminate tension in the body. Then move into your wave breathing, aiming for equal length inhalations and exhalations. Ideally, your body will be as limp as possible, like a rag doll, so there is no resistance in your mind or your body.

When you call your care providers, often they will want to talk directly to mum, as they like to listen to how you're breathing and to assess how the waves are affecting your ability to speak. Let them know you're a hypnobirthing mum, as it's likely you'll be managing the normal stress and intensity of labour much more effectively than they will anticipate, leading some care providers to conclude you're not as far along as you really are. Make sure you monitor the frequency of surges and communicate this to them.

How long will you be in labour?

The answer to this question is: how long is a piece of string? Some people will pop their bambinos out in a few short hours, others will be closer to 40, 50 and sometimes even 60 hours. And I have encountered both ends of the spectrum with students who have used The Calm Birth Method.

The one thing that sets students of The Calm Birth Method aside from non-hypnobirthers is that, even if you're in labour for a LONG time, you can use the tools and techniques you have been practising to help you transform what for some women is a difficult and challenging time into something that helps you feel powerful and in control as you enjoy your labouring experience.

Coping with Distractions

If you're choosing to birth in your own environment, it's much easier to feel comfortable about managing the potential distractions around you. When you know you'll be moving from your home to a birthing centre or a hospital, there are inevitably more distractions you'll encounter along the way, many of which you'll not be able to do anything about. I want to reassure you that outside distractions don't need to be the end of your positive birth experience. You are a master at filtering out distractions in your everyday life and we will be tapping into this awesome skill of yours in the lead-up to and on the big day itself.

The brain processes 400 billion pieces of information every second, of which we are aware of around 2,000. Then we filter out the things we deem unimportant or irrelevant to what is going on in our surroundings. This filtering method stops us from going insane. Thank you, nature. This mechanism is precisely why you don't need to get your knickers in a twist about being distracted. It is also the exact same mechanism I want you to tap into when you're birthing. You can start practising it right now. In fact, you already are, but perhaps aren't fully conscious of it yet.

Your guide for when to tune in to this filter consciously and when not to is really simple. If you're faced with a situation that is creating tension or irritation within you, and you can do something about it (a dripping tap or a snoring partner, for example), then do something about it. Take charge; these situations can be remedied quickly and easily.

If, however, you're faced with a tension-inducing situation that you do not have the power to change, choose instead to focus your attention inwards and start to watch and engage with your breath using The Calm Birth School breathing techniques (surprise, surprise!). Allow the distraction to take you even deeper into a state of relaxation. You can even say to yourself, 'The sound of [*insert distraction*] helps me drift deeper and deeper into a beautiful state of relaxation.' A great time to play with this is when your baby is having a kick and a stretch while you're doing your daily relaxation exercises. Any time there is intrusive noise or something happening within your line of sight, use it as an opportunity to connect to your breath consciously.

This will be easier or harder to do, depending on the distraction, but the more you practise, the easier it will become. A great example is when someone else's child is crying on a plane. Some people will be there wishing they were in the peace and quiet of Business Class, while others will be able to fall asleep amid chaos, appearing completely undisturbed. Your opportunity to practise might come on a bus or a train, while waiting for an appointment or perhaps when you're at a family gathering. The aim, by the time you go into labour, is for you to be a person who can choose exactly when and where you want to focus your attention, and who can breathe through anything.

By making the conscious decision to focus on your breath and allow the distraction to help you go deeper into relaxation, you can change the whole experience of the noises around you with ease. This is a skill that takes practice, but you can do it. The trick is to start early, so I encourage you to make a start today.

Positive power of touch

To help manage the internal distractions – i.e. your waves – take advantage of the positive power of touch.

Touch is an important part of intimacy between couples, as it helps them to relax, and feel connected and safe. All of these emotions are great for producing endorphins and oxytocin, which we now know are the ones we want for a quicker, more comfortable labour (as opposed to cortisol and adrenaline). I recommend that you and your partner make time to tune in to this during pregnancy by enjoying daily massage with each other – and when I say each other, I mean the birth partner massaging mum! Some people will enjoy a firm pressure, but the massage I recommend involves a very light touch.

However, it is also worth noting that, even after taking the time to enjoy this fantastic ritual during pregnancy, when you're labouring you might not want anyone within an inch of you! Some mothers have reported that they needed to be alone, so they went and birthed in the bathroom. That's also absolutely fine. Massage is an extra tool you can call upon should you want it on the day and it will be infinitely more powerful if you take the time during your pregnancy to connect a sense of calm with your partner's touch.

∼ Massage ∼

The light touch of this technique often creates a tingly feeling within the body and the following massage technique is simple to use.

How to do it

Your partner or birth partner uses their fingertips or the backs of the fingernails in a gentle upwards motion, stroking and moving their fingers up your back and across your shoulders, almost creating a T-shape. Repeat and then repeat some more.

You can also bring these soothing strokes up and down the arms, across the chest and nipples, up the neck, behind the ears and into the hair. This is definitely a useful tool to draw upon if labour slows down.

When to do it

A daily 15-minute massage is ideal, as this is not only a great way to build on your opportunity for prenatal bonding, but also the touch of the birth partner becomes a cue or an anchor for you to become even more relaxed every time this type of touch occurs, which can be extremely valuable during labour.

∼ Dial-Down Method ∼

Another great method for distancing yourself from both internal and external distractions is the Dial-Down Method, which is a self-hypnosis tool. This is a great tool to use whenever you notice yourself feeling tense or stressed and you have a little bit of time on your hands. It's the perfect technique to draw upon if you're on public transport or when travelling.

How to do it

Imagine a large dial, with the numbers one to 10 going around the perimeter. As you imagine the dial, see the gauge hovering at number 10. Then take a deep breath in and imagine breathing in calm, and as you exhale, imagine breathing out tension. On the completion of that exhalation, see the gauge move to the number nine. Again, inhale calm and exhale tension, then see the gauge move down to eight.

Do this all the way to zero, and you should notice how much more relaxed your mind and body feel. When you get to zero you can continue to focus your attention on the breath or imagine relaxing in your favourite place. Play with it and follow your gut instinct.

You can amend this technique and visualize a big temperature gauge, a sliding scale of musical notes or a thermometer – whatever makes the most sense to you.

When to do it

Most people start practising this visualization with their eyes closed, but as soon as you become familiar with it, eyes open is equally as effective. If you're ever in the office, particularly if you're in one of those meetings – you know the ones I'm talking about – and need to get out of your head for a bit, this is a great trick to have up your sleeve.

When TCBS mums first start using this technique, it doesn't always come naturally. Many of the techniques I've shared may feel difficult to get your head around at first. Please persevere and be kind to yourself. It is the same when you learn any new skill – it takes time and a lot of repetition to master any new practice to a point where you can do it on autopilot.

Think about when you learned to ride a bike or drive a car, or a toddler trying to find their walking legs. How many times will you see them pull themselves up, only to topple over? It's all part of the learning process, and it takes time, effort and perseverance. It's exactly the same with the skills you're learning throughout this book, which is why I recommend starting now, not a couple of weeks before your guess date!

Time distortion

Hypnobirthing mums often talk about the quickening of the passage of time when they are birthing, and this is a noticeable indication of being in a trance state. It's like being engrossed in a great book – you have just no idea where the time went. You can have the exact same experience during birth, of time working with or against you, depending on how you're feeling in the moment.

Time flies for a woman who feels relaxed and calm and looks forward to each surge as an indication that each sensation brings her one step closer to meeting her baby. Contrast this with a woman who dreads every surge, and wonders when it's finally going to be over and how long labour is going to last. Every minute feels like 10... or more.

When you feel good, your birth partner can capitalize on this sense of time passing more quickly, by suggesting every 20 minutes feels like five. While it might feel silly reading this right now, remember that when you're birthing you're going to be in a trance state and therefore more open to suggestion, so use that in your favour.

➷ **Laura K's birth story** ➷

On Monday morning, we finally got to meet our gorgeous daughter, Imogen Sophia. I was very lucky and had an officially short active labour, but I wanted to share the full story, to show how TCBS got me to that point and how my much longer prelude to labour wouldn't have been so calm without the techniques this course taught me.

I was very anxious about labour from the moment I found out I was pregnant and I started looking into hypnobirthing straight away. I had heard about TCBS indirectly through a social media post by Tom from McFly and signed up for the course when I was 20 weeks.

At first I studied religiously and posted affirmations around my house. I also worked hard on educating my family and fiancé so that they were supportive and not dismissive of my desire to use hypnobirthing. I believe the support from others was key to the course's success for me. Don't get me wrong – my fiancé wasn't begging to watch a video or listen to an MP3 every night (which we did, in fact, do at the end of my pregnancy), but he was on board with the concept and I was so pleased.

After I signed up I did my best to practise where I could, but self-doubt often crept in that I wasn't doing enough. If you're reading this thinking the same thing about yourself, have faith: whatever you're doing will be going in! This was where my fiancé's support was key; he would remind me of the techniques when I had doubts during pregnancy and during labour.

My surges started at 9 p.m. on Friday but had disappeared by the Saturday morning. They returned on Saturday lunchtime but again were gone by Sunday morning, only to return at lunchtime (phew!), at which point they increased in intensity and frequency.

Emotionally, these few days were quite draining. I was constantly thinking, 'Is this going to be it?' and I had plenty of time for self-doubt and panic to set in. However, instead of letting this happen, I watched numerous Calm Birth videos that had been shared through TCBS group (the support you can get from the Facebook group is amazing) and I listened to my MP3s regularly over these few days to keep me calm.

In all my preparation, I had not considered that I might be in early labour for a few days. Each time I felt a surge I would use my breathing (always being reminded by my fiancé to do it) and a random affirmation would pop into my head! I also used a TENS machine to help with the pain relief. Finally, on Sunday morning at about 12:30 a.m. I was having three surges per 10-minute window, so I called the midwife-led unit and they agreed that I should go in.

I was 3cm dilated at 1 a.m. so they gave me two hours before they checked me again, offering me pain relief tablets, which I declined. By 4 a.m. I was at 4cm and my options for pain relief opened up. During the waiting to get from 3cm to 4cm and making decisions about how my labour progressed, I reverted to the BRAIN acronym. I felt empowered to be making those decisions and not sheep dipped into having a medically managed birth.

Once I was at 4cm I moved onto gas and air then had a wonderful aromatherapy massage from the midwife. Things progressed pretty quickly from there. I went from 4cm to delivering my baby in the water within three hours with just 20 minutes of pushing!

I have to be super honest here and say I was not silent or serene like some of the incredible women in The Calm Birth videos, but I was incredibly empowered and I felt OK being very primal while I gave birth. I learned that that was more than OK and it clearly resonated with me on the day!

Overall, my biggest credit to TCBS would be keeping me calm while in early labour and throughout my pregnancy. It educates you to be a woman who is fully in control of the decisions you make. Your labour may not go the way you hoped and you might have limited choices, but you do have choices all the same and you should feel comfortable in making them without fear. Thanks for the education, Suzy, and for helping me to have a genuinely positive and empowering birth experience. I couldn't have done it without you.

With love and gratitude from me, my fiancé and our precious little girl.

Chapter 16

What Happens Once You're with Your Care Provider?

Once there is space for you on the labour ward or in your room – if you're not birthing at home – most of the time you'll be offered a vaginal examination. Note the use of the word offered. A vaginal examination, as with all procedures offered during birth, is based on informed consent, so it is your right to decline a vaginal examination should you wish to.

Why do some women decline this? Well, some mums-to-be want to have a completely undisturbed birth and believe baby will arrive whenever he or she is ready. Their perspective is that a vaginal examination does nothing to aid that process. It's also recognized that a VE only measures a moment in time. So if you go to the hospital, are examined and a well-meaning midwife says, 'Oh, you're only 2cm,' the use of the word 'only', plus the fact that you're 2cm, can be hugely deflating. If you do opt for an examination, remember that it is just a moment in time – labour isn't linear and things can progress very quickly.

I can't emphasize this enough: if you opt for a vaginal examination – and many women do – and you're not progressing as quickly as you would like, relax and do as many oxytocin- and endorphin-producing activities as possible.

This really is your birth partner's time to shine and remind you, you birthing beauty, that this is just a moment in time and that you're doing amazingly. Just because it's taken 10 hours to reach 3cm, doesn't mean it's going to take another 10 hours to get to 6cm. When a mother is relaxed and birthing as actively as possible, it is completely possible and not uncommon for mum to go from 3cm to 8cm in an hour!

If you find that you haven't progressed at the rate you would have liked, know that it's totally normal to feel frustrated and concerned about how much longer things are going to take, but then consciously choose to release the expectation or the wish for things to be different. Let go of any tension, doubt or anxiety that may be trying to creep in and simply be accepting of the moment.

This is not easy to do if you haven't been practising breathing and sending relaxation to the different parts of your body, both at home and in less conducive places for relaxation, such as on public transport. So make sure you practise in advance.

The slowdown

Whether you're travelling to a hospital or birthing at home, it's very common at one point or another during your labour for things to slow down. Depending on what stage you decide to leave home, if you do at all, a slowdown occurs because you're leaving your cosy and familiar home environment. This is normal and all part of the evolutionary armoury that aims to keep us and our babies safe from harm while our subconscious assesses the reasons we have moved and whether it feels good to continue with the labour.

If your body and baby decide to take a little break, you might like to use one or more of the following ideas to help get your surges going again:

1. **Breathing techniques:** You'll have practised these so many times by this point that your muscle memory associates breathing with feeling good, being calm and at ease, which acts as a signal for you to produce more endorphins and oxytocin.

2. **Touch:** Remember the power of positive touch, whether it is a little bit of massage, stroking or holding. If you feel open to being held or touched, give your birth partner the green light, if they have not already taken their cue.

3. **Walking:** Walk upstairs while leaning forwards to help baby put more pressure on your cervix. Walking sideways upstairs can also help – think of a crab!

4. **Visualization:** See your baby moving down the birth path and how good it's going to feel to be holding them in your arms, or listen to your Birth Rehearsal MP3 – all roads lead to oxytocin.

5. **Nipple stimulation:** This is great if things slow down as the stimulation of the nipples impacts the body in a similar way to when a baby is suckling, causing the

body to produce oxytocin to help promote bonding and attachment. Of course, oxytocin production will increase the efficiency of your labour.

6. **Laughter:** Watch anything that makes you laugh. Download it onto your tablet beforehand so you've got easy access to it. We laugh when we're relaxed and when we're relaxed we birth babies more quickly, easily and comfortably. Trust that your body and your baby know what to do and things will pick up again at exactly the right time. The best thing you can do is relax.

7. **A warm bath:** Water can either speed things up for a woman in labour or slow things down, so it's wise to be mindful. If you're experiencing irregular surges that slow down after you have hit the water, get out. However, the flipside of that is the lovely warm water stimulating your oxytocin and endorphins, and helping things along beautifully.

Best avoided

While I think films (hilarious ones) are a great idea, mobile phones, laptops or anything that will engage the rational, analytical part of your brain should be banned during birth if you're looking to create a quicker and more comfortable birthing experience.

And certainly no social media once you're in active labour: no tweeting, Facebook or texting your friends. The more you disengage from the left side of your brain – the part responsible for critical analysis – the easier you'll find it to connect with the sensations in your body, rather than analyse them from a cognitive distance.

It's also worth thinking about whether you want to let anyone know once things have started to get going, because if you're going to experience a longer labour, the pressure of having people texting you or your birth partner to find out what is happening can also create unnecessary stress, pressure or analytical engagement.

Working with your care provider during labour

Ensure you have at least two copies of your birth preferences with you, so if there is a change of shift when you're birthing, your new care provider can get up to speed.

As we have discussed, who you have in the birthing room is hugely important and your birth partner should be responsible for managing your environment, so highlight this section for your birth partner to read.

It is the birth partner's role to be fully tuned in to mum, so they can see how she is feeling and talk to the people who are working for you. Please remember, they are working for you. If there is any hint your practitioners are not supportive, or there is a personality clash, your partner needs to take the lead and in a calm but firm way ask for you to be cared for by someone else. It is not your responsibility to be concerned about how that is managed, but it is your partner's responsibility to hold and protect the space, so that you have an optimal birthing environment, which is full of peace, love and support. Then you'll be able to make the decisions that are right for you and your family.

Remember, you'll only have this experience once. So you have to make the choice about whether it's best to be more concerned with ensuring that you create the most positive birth experience for you and your family, or worrying about potentially offending someone who is unable to support you in the way that you need in that moment... I know what I'd choose.

Decide beforehand whether you would like your birth partner to act as a go-between for you and your care providers when it comes to making decisions about the birth, or whether you would like to take the lead. As always, there are no right or wrongs when it comes to this, but if you're looking to birth more comfortably and quickly, the less you engage in conversation, the better.

Many hospitals will want to see progression of about 1cm dilation per hour. But women are not machines, so we often do not work like this. Once again it's important for your birth partner to be tuned into how you're feeling so they can alleviate any pressure you may encounter from people trying to speed things up.

If you find yourself in a situation where people are trying to hurry things along, use the BRAIN system and ask very specific questions. This will provide you with specific answers, and allow you to make informed and personal choices about the course of action that is right for you. Always bring any suggestions about proceeding back to the indications at that moment and be as specific as you can.

Example questions

- What specific evidence suggests that there will be a risk of [*insert complication*] happening?

- What specific signs right now indicate that waiting for the next 30 minutes would be harmful to mum or baby?'

- If we choose to wait another 45 minutes, what are the possible risks, and what evidence do you have that those risks are applicable to my baby and me?

Language

By this point I hope you're being mindful of the language you're using (*see pages 20–21*) to think and talk about birth. I've talked at length throughout this book about using different terminology and being very mindful of the kind of thoughts you think. Unfortunately, you can't control the words used by others. That said, it wouldn't do you any harm to include how you would like your care providers to talk to you in terms of comfort levels and waves, as opposed to pain and contractions, in your birth preferences.

- Certain words and phrases used by the medical profession can be upsetting or even distressing to hear during labour. Prepare yourself in advance by making a note of the terms below, and making a decision about what you and your birth partner will do if you notice this language during your birth.

- For example, if someone in your space is using negative language, it is totally acceptable for the birth partner to respectfully ask the care provider to speak with them directly first, quietly and even outside the room if possible. The birthing environment should be kept as emotionally safe and relaxed as possible.

Some common phrases that are often used are as follows:

- **Failure to progress:** If you happen to hear this when you're labouring, remember that you always have options about whether or not you would like to follow the lead of your body, or if you would appreciate some assistance. Once again, go back to your BRAIN and assess what you feel is going to be right for you.

- **Incompetent contractions:** Whoa! Incompetent is definitely another phrase that should be banned from birthing rooms. There is nothing incompetent about your body and sometimes, for good reasons, your baby will stop moving down the birth path and you'll require assistance. It might simply mean that you need to move around, change position and take the pressure off to allow the oxytocin to do the job it was designed to do.

- **You're only [*insert number*] centimetres dilated:** This is a HUGE no-no in my book. Dilation ONLY measures a moment in time. Just because it took you three hours to get to 4cm, it doesn't mean it's going to take you another three hours to get to 8cm. Please include in your birth preferences that if you choose to ask for an update, your care providers should frame your progress with encouragement, like 'You're doing well; you're currently Xcm.' Or 'This is where we are at right now.'

- **How much pain are you in?** While we don't ban the P-word, it's much more helpful for your care team to ask, 'How comfortable are you?'

- **You're not allowed:** It doesn't matter what this is in reference to, quite frankly it's not true. It's your baby and your body, so if you want something to happen you're entitled to have it happen.

- **We are going to:** Informed consent is required for all interventions, treatments and procedures, so always feel entitled to ask questions if someone looks to be moving forwards with something you don't understand or agree with.

With any decisions you make about how involved you would like your care providers to be, remember that you have choices.

If you opt for support to move things along a little, and make this decision while feeling calm and in control, you remove any ambiguity because you can ask specific questions and receive specific answers. This contributes hugely to you being able to look back on your experience, knowing you had the right birth for you, regardless of how your baby entered the world. That can only be a positive thing.

Things to be mindful of during the active phase include:

- Changing positions and keeping as active as possible.

- Using positive touch and words for comfort and reassurance.

- Making your birthing environment – particularly if you're not birthing at home – as lovely and as nest-like as possible.

- Having your birth partner hold the birthing space so it feels calm, private and safe.

Chapter 17

Second-Stage Labour

You are considered to be fully dilated at 10cm, sometimes referred to as the down phase or second stage of labour, and it means your baby is close to being ready to emerge. There are three main factors to be aware of, which I'll talk you through now.

The three signs of second-stage labour

For many, the second stage of labour can mean an increase in intensity, although this is not always the case.

A pause

At this point, just as during early labour, the body may take a natural break as your body and your baby instinctively prepare for this new phase. This is not the same as the slowdown (*see pages 158–159*) and it's nothing to worry about or hurry, as long as your baby's heartbeat is strong and consistent. Sometimes this pause will be 10 minutes, and I've even heard stories of the break lasting a few hours.

If you're birthing at home with a private midwife, it is far more likely they will feel comfortable going with the body's lead at this point. If you're birthing in a more medicalized environment, it's more likely that your care team will want to move things along if the break is deemed to be too long. If you're faced with this scenario, remember to use the BRAIN system to determine what course of action is going to be best for you and your family, and to make sure you're clear as to why any suggestions are being made (*see also pages 74–75*).

Sometimes instead of a pause, you might notice an increase of intensity or frequency in the waves you're experiencing, or perhaps a change in the sensation of the surges. Everyone's experience is different.

Pooing

Many women fear emptying their bowels while bearing down in labour, and sh*t does happen (sometimes), but remember that it's all for our own good. Once you move into the second phase, the body's primary focus is to work as efficiently as it can and so it expels anything that may inhibit the baby from moving along the birth path. As the baby moves past the bowel, any waste is emptied to leave the birth path as clear as possible and thus make the baby's descent easier.

So should you start to feel an overwhelming urge to poo, be happy, because it means that you're in the second stage and close to meeting your baby. Sometimes you might feel the urge but not end up doing so. This is great, too. It simply means that baby is moving down the birth path but that you don't have any waste to expel.

If you feel the need to poo once you're fully dilated, resist the urge to jump up and go to the toilet if you can, and move your attention back to your breath. It's time to begin breathing your baby down. If you're not participating in vaginal examinations, trust yourself to know that things are changing and you're moving into the second phase. Some people may call this the 'pushing phase'.

For some women it will feel as though the body is taking over and for others it will be a more conscious decision to change the way they are breathing with their surges. At this point I invite you to choose to focus your energy and your breath downwards, working with your body and assisting your baby's journey.

Feeling nauseous

Another great indication you're moving into the second phase is vomiting or feeling nauseous. You could find yourself projectile vomiting across the room after a substantial period of labour, accompanied by an increase in either the frequency or intensity of your surges, do not worry. Alternatively, you may notice your surges dying off as your body pauses before you prepare to cross the final hurdle (there's no one direct route). It's all positive!

Pushing and breathing

When we watch people giving birth on TV, the second stage is what we usually associate with lots of forced pushing, and care providers shouting and coaching the birthing mother. Some of you may want the support of your midwives, and others may prefer quiet to let the body do what it innately knows how to do. As always, nature thinks of everything and, as with every other birthing mammal, there is no need for most human mothers to force their baby out. The body has a unique mechanism for doing this on our behalf, called the natural expulsive reflex. You'll already be familiar with it from using it every day when you go to the toilet to empty your bowels.

Breathing baby down

When you're in this downwards phase, the best thing you can do to aid this process is imagine sending your breath and energy down into the ground, past your baby, around your uterus and into the floor. This allows your body, pelvis and vagina to be as relaxed and open as possible for a quicker and more comfortable birth.

Some women experience an overwhelming urge to push or bear down. If this is you, I advise you to go for it. Listen to your body; do not resist it, even if 15 minutes previously you were told you were 'only' 6cm dilated. As I said earlier, vaginal examinations only measure a moment in time and things can change very quickly. Trust your body.

So what's the difference between pushing and bearing down if your body is telling you to do so, and forced or coached pushing? In a word: tension.

What I'd like you to do, either when you're sitting on the lavatory or sitting in your chair right now, is to pretend that you're emptying your bowels. What do you notice? You should feel the muscles around your sphincter contract and tighten – which is the exact opposite of the action we are looking for with a calm and positive birth for both you and your baby. When you can breathe your baby down and work with your body, your baby tends to make their entrance in a calmer, less explosive way, which also helps to keep the perineum intact. Bonus.

Having said that, for a multitude of reasons, some women opt to force-push their babies into the world. As with every piece of advice or insight I offer throughout this book, if this is a conscious decision, coming from you, then it's all good. If you're faced with a situation that dictates that forced pushing is the best way forwards and it feels right for you, then it's right. I really want you to hear me on this: you can't get this wrong. Just go with what feels right for you on the day.

∼ Birth breathing ∼

Use this technique when you're experiencing a wave or surge when you're fully dilated.

How to do it

The best way to aid the natural expulsive reflex is to work with the breath in a way similar to when you're wave breathing (*see page 46*). The main difference is that you place all the emphasis on the exhalation; the out-breath needs to be very long and very deep. It can be useful to use a visualization to accompany the out-breath – anything that reminds you of the importance of staying open, relaxed and moving downwards.

Some women use the words 'open, relaxed or release,' while others will think about there being no resistance, or imagine a flower opening, or will picture something significant to them that helps keep the idea of openness in their mind.

It can be really helpful to work with noise when you're getting to this stage. Although some will feel equally comfortable working with the breath alone, others will want to hum, shout, groan or even moo. If it hasn't already, it can get incredibly primal at this stage. This is nothing to be fearful of for either you or your birth partner. No resistance is the main aim of the game. If you want to howl, just howl! Whatever you instinctively want to do is all good – seriously. The only thing to be mindful of is to use the noise and the energy to send your power back down to your baby and your uterus so they can finish the job.

When to do it

The best place to practise this technique is when you're having a poo. If you're at home, hum when you're on the loo so you start to feel more comfortable and familiar with directing your sound and energy down in that way.

This is great if you're suffering with constipation, too. It won't shift everything immediately, but by applying patience, the humming and breathing will see your natural expulsive reflex start to get things moving much more quickly and comfortably.

Chapter 18

Transition and Crowning

There often comes a moment during birth when a woman thinks, 'I can't do this any more!'

The second phase usually lasts a maximum of two hours – although this will vary from woman to woman. Sometimes, right before baby is about to emerge, a woman may feel overwhelmed, like she can't do it any more and she wants to give up. This is called the transition stage. If this happens to you, this is an amazing sign for you and your birth partner that you're now within spitting distance (excuse the analogy) of meeting your baby.

This is the time for your birth partner to remind you of what an amazing woman you are and let you know you have reached the final hurdle. This period of wanting to give up or feeling like it's all too much right before the prize doesn't happen to everyone. However, if it does happen, take heart in the knowledge it is only likely to last for around 15 minutes and you really are going to be meeting your baby any moment.

Once transition has passed (should you experience it at all), the last moments of birth are all about you working with your body to breathe baby into the world. Just as baby is about to emerge, some women experience an intense tingling sensation. This is your baby stretching and pulling on the vaginal walls. The intensity only lasts for a short time and is often followed by numbness, so aim to stay relaxed. Sometimes the sensation of your baby's head can come as a shock – as crazy as it may sound, don't be tempted to close your legs (it does happen). Focus on your birth breathing and waiting for the next surge to breathe out your baby's head fully, which should then be followed by their body, in just two to three surges.

∂ Charlotte's birth story ∂

Throughout my pregnancy I struggled with various physical and mental stresses and pains. I had numerous hospital visits with kidney stones, urine infections and bleeding which, when accompanied by running two businesses, working 15-hour days seven days a week, plus teaching dance classes and renovating a house, made me only too pleased to take a day off work one Friday. I enjoyed a mum-to-be spa day that was a baby shower gift from my lovely friends, and I treated my mum to treatments too, as she had been a fantastic support to me throughout my pregnancy.

We had watched all The Calm Birth School videos together and discussed how I was going to have an ultra-calm and happy birthing experience. I had been having cramping for about six days and she would remind me to remove my frown and stay positive. So after a relaxing day off work and a three-hour afternoon nap, I came home feeling really calm and relaxed. That evening at 11 p.m., as I sat on the sofa with my partner, Steffen, my waters released. I was so excited, bouncing around with so much joy singing, 'We're having a baby! Baby is coming; he's coming so soon!'

I think it was this happiness that set off the endorphins and oxytocin, and things went from zero to 60 from that point. Within 30 minutes, the surges had started, but very soon I realized that there weren't any breaks in between them.

For about an hour I struggled to move, get dressed or talk. I knew I was in established labour: the feelings were very intense and painful, and honestly, there was a point where I thought, 'I just can't do this.' I knew that I needed to get to the hospital; maybe then I could get some pain relief.

The birthing centre that we had chosen was closed due to staffing shortages, so we made a quick change of plan and went straight to the local hospital. When we arrived I was wheeled in, because by this point I couldn't even walk. I knew I was close to pushing, but because I was silent with my eyes shut the midwives didn't believe I was in established labour and left me in a waiting room, with horrible bright lighting.

I was determined to stay strong, positive and, above all, calm! Finally, I was examined and the midwife was shocked to be able to see my baby's head! With no time left to fill up a birthing pool as planned, I was rushed to a labour room, where I got on all fours and began to push. I told Steffen to put my Calm Birth School playlist on and get the lights down low. I focused on the positive affirmations while Mum and Steffen took it in turns to rub my middle back, which was hurting a lot.

I would highly recommend essential oils like clary sage, which really helped to speed things up while relaxing me. It took about 30–40 minutes for me to breathe my baby out, literally without any pain relief or even a sound. It was an amazing experience, and being really positive and fearless about giving birth is what I think helped me the most: calm mind, calm body, calm baby, calm birth!

It is so important to remember this and I can't thank TCBS enough. Meeting you and learning from you helped to make my birthing experience a truly magical and special event, and it is now my job to spread the message that giving birth can be amazing and very positive.

Chapter 19

Your Baby's Here!

Even though many a wise parent will tell you that the real work begins once baby arrives, there's a little bit in between giving birth and parenting starting that is worth thinking about and preparing for.

Those moments where you hold your baby for the first time are not only magical because you're getting to look into each other's eyes, but also because you start to close the complex physiological and emotional loop of birth.

Cleaning your baby

If your baby is born on dry land, as opposed to in the water you'll notice a white coating on your baby's skin. This is called 'vernix'. Some mothers prefer their baby to be completely wiped down before enjoying skin-to-skin contact with their child; others prefer things to be left au naturel. The choice is up to you. However, some people argue that babies are not born dirty and the rush to clean them just isn't necessary; vernix acts as an antiseptic moisturizer for baby's skin and protects them from a whole host of infections.

Delayed cord clamping

Delayed cord clamping is when the umbilical cord is not immediately cut. Some parents are happy to leave it for two minutes before cutting, while others prefer to wait until the cord has stopped pulsating completely before cutting so they know baby has everything they need from the placenta.

One of the things you can research is whether delayed cord clamping will be suitable for you or not. As you've read in some of the stories from TCBS mums throughout this book, the idea of delayed cord clamping might seem perfect for you, until you actually go into labour, when circumstances can change very quickly.

If this is of interest, please do your own research into the benefits and drawbacks of this course of action. Be mindful that your choice is very important to include on your birth preferences sheet.

Birthing your placenta

Birthing your placenta is one of those things that for some women can come as a total shock (*ahem, that would be me*). We spend all of our time preparing to stay calm, relaxed and in control for our babies. Once we finally get to welcome them into the world, it can sometimes feel like a bit of a surprise to realize the job isn't over yet. If you have been giving birth in a pool, you'll need to get out and birth your placenta on dry land. And while some placentas plop out without mama giving it a second thought, sometimes it takes a little bit of effort. If you're at the end of the spectrum that requires a little bit of focus, remember that all the skills you applied to birth your baby can be just as relevant for this third and final stage.

It is worth researching whether you would like to have a managed third stage when birthing your placenta, or whether you would prefer to let the process take its natural course. If you have a managed third stage, you'll receive an injection of synthetic oxytocin to stimulate the uterus into surging. This often means the third stage is over within half an hour. A managed third stage is often recommended, as it tends to reduce the number of women experiencing a large loss of blood after birthing. One of the disadvantages of taking a synthetic drug to speed up the process is that it increases the chances of retaining parts of your placenta,[20-21] which can lead to infection and sickness.[22] Once again, do your own research so you can decide which option is right for you.

Skin to skin

The impact of skin to skin is one of nature's most awesome design features, and creating the space to allow this to happen immediately after birth has a long list of benefits! So barring special circumstances, this is a fantastic way to welcome your baby into the world.

Skin-to-skin contact with your baby stimulates oxytocin, another amazing ingredient of the birthing process. It turbocharges the bonding process, helping you to want to protect and nurture your baby immediately. However, it is important to say, even with the oodles of oxytocin coursing around their veins, some mums take a little while to connect with their baby fully, and if this is you, that's OK and a normal response, too.

The oxytocin not only helps with attachment bonding, but it also begins stimulating the uterus, so that within the hour you'll have birthed your placenta.

When your baby lies directly on your bare chest – and on your partner's, too – your baby's skin starts to colonize with your friendly bacteria, which is amazing because it means alongside the antibodies in your milk (if you're choosing to breastfeed), you begin protecting your little one from any unsavoury bugs and germs that may be lurking.

One of the byproducts of oxytocin production is a warm, fuzzy feeling. The warmth helps to start regulating baby's body temperature and reduces their cortisol levels (the stress hormone). It also triggers your milk ducts so that you can begin feeding baby right away.

If, owing to special circumstances, you're unable to move forwards with immediate skin to skin, but it's something you don't want to miss out on – particularly if your baby is preterm – you can always ask your care providers about Kangaroo Care. This is where mum gets to carry baby around on her chest like a little joey, which again helps you both to form an attachment, aids breastfeeding and importantly colonizes baby with your friendly bacteria to help protect against illness and infection.

Breastfeeding

Breastfeeding, as with birth, is an individual journey that for some women comes very naturally and for others can be extremely challenging.

While I don't go into any details about the practicalities in this book, please do check out www.thecalmbirthschool. com/courses for the 'Infant Feeding Workshop'. What I would like to touch upon is the emotional side of feeding because, next to lack of sleep, it's often one of the things that most takes new mums by surprise. I often say that it's not until you have your own child that you appreciate why so many mothers think of themselves as 'feeders'.

Ensuring your child has enough nourishment is solely down to you if you choose to breastfeed. If you find that everything does not fall into place naturally around breastfeeding, the feelings of guilt and shame at not being able to feed your baby in the way you had envisaged can be overwhelming.

The evidence shows that statistically, women who seek out support and guidance before their baby arrives tend to have a more positive breastfeeding experience. Women who look for assistance from a lactation consultant as early as possible if things are not going as well as anticipated are far more likely to breastfeed for longer. Don't suffer in silence or wait for six weeks before acknowledging that help would be good. Needing help does not make you a failure.

If, after seeking out support, you still feel as though breastfeeding isn't for you, please do not beat yourself up. The happiest children have happy mothers. Sometimes, for a multitude of reasons, you may choose to stop breastfeeding. Please hear that this doesn't make you a bad person or a bad mother. Educate and empower yourself, and remember it's your body, your baby, your decision and nobody else's business.

Conclusion:
Moving Forwards

I hope as you reach the end of this book a quiet confidence has been awakened within you, or that perhaps the inner strength you have been silently drawing upon now has its own voice, which you feel very comfortable sharing with the world.

Hypnobirthing isn't about producing perfect textbook births. If only it were that easy. However when, as a woman, you understand what you can control and you let go of what you can't, you can step into creating the most positive birth experience, regardless of the way your baby chooses to enter the world.

This is your body, your baby and your unique experience. You are a warrior woman, a goddess who was born to give birth. You now know that birth is a natural and normal event, not a medical procedure.

You know that being able to embrace and accept what is going on in your body, rather than resist it, can help you to manage the sensations of birth far more effectively.

You understand that, although your emotional state will have a huge impact on the way you experience birth, the birth environment, your care providers and your birth partner also have extremely important roles, and they should all be there to support you fully. You have the right to ask for what you both want and need.

You now have a toolkit of specific techniques that will help you to remain calm and at ease throughout your labour and birth.

So what's next?

No one can force you to do anything you don't want to do, but in order to get the most out of this book, I suggest that you go right back to the beginning and read it again, highlighting all the salient points. Yes, seriously. Remember everything I've said about preparation. Never underestimate its value.

There really is no secret to giving birth, other than to be present in any given moment and to accept what is going on with your body. The nub is, for most of us out there, that it takes time and practice to be able to tap into the acceptance that you need on the day. So hear me when I say that I believe in you and you have got this, and make sure you put the time in, do your practice and watch how you can handle whatever birth sends your way with ease.

Next, set dedicated time aside in your calendar to practise the breathing techniques and visualizations. Don't simply set the intention to practise: by putting appointments into your calendar, you'll make the appointments with yourself 'real' and will be much more likely to show up for them – especially if you've got a full life already. Remember, you can download The Calm Birth Method practice guide at: www.thecalmbirthschool.com/bonuses

In closing, I want you to know it's not all about birth for me, it's about life. This is about *you*. I want you to know that you count and your voice is important.

While giving birth should never define a woman, the process of fully connecting with your inner voice, strength and wisdom you learn to exercise, stretch, pull and embody both during your pregnancy and your birth will provide you with an opportunity to celebrate and own your magnificence as a woman. You are not only a human being who is a living, breathing, walking miracle herself, but also someone who creates miracles.

As you gain confidence in communicating your needs, wishes and desires for yourself and your unborn baby, you begin to understand the power you contain is the power needed to conquer worlds, whether these worlds are within the private sanctuary of your home or breaking down glass ceilings in the boardroom. This path (while not the only way) is a path to your power in life.

It starts with knowing what you want and telling other people, then looking your fear in the face. Stare at it head-on and see it melt away as you trust in yourself, your body and your baby and lean into your strength.

When you can navigate your birth in a way that leaves you feeling like a lioness, it changes things. It changes you. You know you can do anything.

You can do ANYTHING.

You have got this!

All the love

Suzy xo

Postscript

Another birth story? Oh, go on then, have two!

⤜ Hayley's birth story ⤛

Today is my due date. However, six days ago our little Huxley Isaac decided to come into the world.

My waters broke at 11:30 p.m. on Sunday. I called the midwife and she suggested going back to bed as it could be a while before anything happened. (I'd been having really strong Braxton Hicks for a week.) I managed to doze off until about 4:30 a.m. then decided to get up; the surges were mild and 10 minutes apart, so I called my mum at 5:30 a.m. to come over to help get our four-year-old to school.

I then spent the next few hours trying to doze off, walking around the house and having small snacks, all the while welcoming the surges. I found walking and standing up was the only way to cope with them. We had lavender burning, candles and plinky-plonk music, as my other half calls it! It was super relaxing; my mum said she felt as though she were in a spa!

I walked to the end of the garden and back again a few times, and the surges (when we timed them, which wasn't all the time) ranged from four minutes to eight or nine minutes apart.

We called the midwives again at 11:30 a.m. They told me just to relax, not do anything, let my body do its thing and, if I wanted, to have a bath. I had a bit of an ache down below but not a pushing feeling, so one midwife advised me to try to feel myself in the bath to see what stage I was at. After getting into the bath, and feeling what I thought might be the baby's head, I had three close strong surges. Luckily the midwives were already en route.

The next couple of hours went so fast. I had the affirmations on while I rocked holding on to my partner or my mum. I also found 'ahhh-ing' sometimes on the out-breath of wave breathing helped. The pool was only just filled in time; I got in at 3:18 p.m. and only three surges later at 3:34 p.m. our baby boy arrived!

I'm so thrilled with our calm, positive birth experience, which despite being definitely uncomfortable at times towards the end, was also, in a way, euphoric. This birth was 16 hours from start to finish, compared to 30 hours with my first. I only had three surges where I let go and birthed him, compared to two and half hours of forced pushing with my first. We were in the comfort of our home and tucked up in bed within an hour!

Thank you to The Calm Birth School for everything I've learned, for changing my perception of birth – as well my mum's and hopefully others' too – and all the support I've received from the Facebook group.

∽

❧ My final birth story – welcoming ❧ Aluna Grace Ashworth into the world

My body tends to like my babies well done. Caesar was born 11 days after my guess date, Coco was born 13 days after hers and, following what felt like an exceedingly long wait, Aluna was finally born at home in the water, delivered by me – 15 days after her guess date. And it was everything I could have hoped for.

The day before I went into labour started really positively. I visited a new reflexologist and, I'm not going to lie, there was part of me that was hoping the treatment was going to shimmy things along. But as I say frequently during The Calm Birth School video course: 'Natural methods of induction are only useful if baby is on the verge of coming. If baby is not, nothing's going to work.'

Later that day my lovely midwife Virginia Howes came to see me. A couple of days before, I had told her if nothing had happened by Monday, I wanted to have a sweep. So when Virginia arrived she checked for baby's heartbeat and then asked if I still wanted a sweep. When I said yes, being pretty no-nonsense – which I'd loved right throughout my care – Virginia told me what a 'good' sweep would involve and it didn't sound nice. I'd had sweeps with both of my previous pregnancies and, while they were slightly uncomfortable (and ineffective at kick-starting labour), I'd gone with them as the lesser evil over a potentially having to have an induction. With this pregnancy I'd opted to have an independent midwife and so the only conversations I'd had about induction were with other mums at the school gates.

Both Virginia and Lauren Derret, my doulas from The Whole Nine Months, were completely supportive and clear that induction was not on the agenda for me. The baby had been crazily active throughout my pregnancy and, while she was still moving around like a mad thing and my blood pressure was as it should be, there didn't seem to be any reason why I should go in for an induction – particularly with my past history of carrying late.

However, I was now on day 14 past my guess date and, although I felt great after my reflexology, there was part of me that felt I'd done pretty well in terms of staying calm and mostly positive, and if I wanted a sweep I wasn't letting myself or anyone down. Despite that, the thought of Virginia rooting around my nether regions for five minutes trying to stimulate my uterus wasn't appealing, and so before we went down that avenue I asked for a vaginal examination (the only VE I had during my entire pregnancy and birth).

It was uncomfortable and when Virginia reported back she didn't give me the feedback I wanted to hear. I knew it wouldn't stay that way and I remembered my own advice about it only being a 'moment in time', but it still took a while to get my head round the idea that it wasn't my time yet.

I texted Lauren to let her know and she immediately sent back some reassuring words – even though I didn't appreciate them at the time. But that's the beauty of having a doula – you don't have to reassure them; they are there for you and your feelings 100 per cent. Anyway, I decided to bunker down, on went Netflix and I watched films back to back until about 11 p.m. when I switched on Billions. *Not the type of programme you watch when you're looking for an oxytocin fix, but what the heck, I wasn't going to go into labour that night anyway. During* Billions, *the baby started going crazy; my belly was like her own personal punch bag, it was really full-on and about 15 minutes before it ended I noticed that I was feeling quite damp. I'd been losing quite a lot of discharge (apologies if that's TMI) and so I didn't really think too much of it. Then I could feel myself losing more water and thought,* hang on a minute. *I stood up and there was a wet patch on the sofa. Woohoo! My waters had released. (I have a leather sofa, so no damage was done.)*

I was super excited, because my waters hadn't released outside active labour before, so I really wasn't expecting it. Of course, as soon as I stood up, there was no question about what was happening. It wasn't a big gush but enough to make me hop, half run (really, in my condition) to the bathroom so I could sit on the loo, check that the waters were clear and grab a towel. By this time it was midnight. So all excited, it crossed my mind that I was already pretty tired (I should have been in bed by 10 p.m.) and I wondered if things really were going to kick off that night. As I got into bed, my husband opened one eye and I told him my waters had released, and he looked like he might get up for about half a second, before I said it was fine, and of course half a second later he was completely asleep again – #men!

As I relaxed in bed the first thing I reached for was my phone. I texted Virginia and Lauren to let them know the good news and then on went the MP3s. First, I put on the Birth Rehearsal, which had had me feeling super emotional in the lead-up to birth, and then the Fear Release... It was important that I listened to that, as even though I was ready, there had been lots of points in the pregnancy where I had questioned whether I was doing enough practice and I also felt the pressure of needing to have a great birth. I wanted to let those concerns go so that I could just concentrate on the job in hand. I was hoping that I might be able to go to sleep, but what I had thought were pre-labour surges (and in fact were actual surges) were too intense to sleep through. So I just listened to the MP3s sat up slightly in bed and focused on my breathing. I did this until just before 3 a.m., when I thought I would try to lie down on my side... Immediately everything felt super intense and I realized I needed the toilet. Again, I was really pleased I needed to poo and it was pretty loose (I'm doing it again aren't I? TMI), so I decided to get the hubby up and ask him to fill the pool. Jerome got up straight away and I got busy texting Virginia and Lauren.

I'd been timing my surges for a while and I was experiencing three surges in 10 minutes, ranging anything from 40 to 60 seconds each. So when Virginia asked me if I wanted her to come, I hesitated before getting back to her, because I didn't want her to come out at 3 a.m. if she didn't have to. But given she was an hour's drive away, I thought it would probably be best if she did make her way over, because who knew what would be happening at 4 a.m. and I had been pretty quick with Coco. I then texted Lauren and took up my position in the living room. I had my laptop by my side, playing a slide show of the family photoshoot that the lovely Philippa James had taken a few weeks before and TCBS MP3 playlist was on a loop. I had chosen the Powerball and the Birth Rehearsal affirmations to listen to throughout the birth. Instinctively I sat down on my knees facing the sofa, with my arms resting on the seat, supporting my head. I had stopped timing by this stage and whenever I felt a surge I circled my hips. Everything felt really manageable. I couldn't have told you how far along I was but, because I felt very calm and in control, if I had to guess I would have thought I was still quite early on.

On reflection, time passed really quickly, because before I knew it Virginia was there by my side watching me wind my hips around a surge. She asked me if my surges were any more intense, but I couldn't give an honest answer, because I didn't know. I just knew that it was all quite manageable. As the surge passed I think Virginia had a listen to baby and all was well. I hung out for a short time longer in the dark of the living room before moving into the kitchen, which was all set beautifully for me by Jerome. There were candles running along the countertops, our fairy-light wreath hanging by the door, and affirmations pinned up along the window and the wall. It was perfect. And when I moved into the kitchen, he brought the laptop so I could continue listening to the playlist on low.

Even though the pool was prepared, I wasn't quite ready to get in. So I resumed the position I had been in, in the living room on my knees, leaning on the seat of a chair. I think Virginia must have sensed things were moving along, as she asked me if I was ready to get into the pool. I think I said no to start with and then a little time later she reminded me I could get into the pool if I wanted. I remember feeling that I might as well, but as soon as I did, I felt amazing about getting into the water, yet also more aware of the surges. The frequency of the surges immediately increased and I had to bring all my awareness to reciting my affirmations. I smile when I think about what I was reciting to myself on the day. The one that sticks out was, 'I am a goddess.' I also said many times, 'I can do this.' And I had to remind myself to release all the tension from my body.

I remember leaning over the pool (same position: on my knees, leaning forwards over the edge) and being aware that it was getting lighter, and looking out over the garden thinking how beautiful it looked. Then I felt like I needed to wee. I told Virginia that I wanted to get out to go to the loo. My surges were coming thick and fast at this point and she told me I could go in the pool, but I wasn't up for that. It's really interesting that I was completely present in between my surges this time, which was very different to Coco's birth, where I stayed 'in the zone' throughout.

Jerome had been quietly watching and supporting from the side of the pool at this point, and Virginia asked him to help me out before the next surge arrived. Too late – the next one came! I just managed to get out of the pool when the next one came and it was really intense. Maybe this is partly because I had got out of the water, but also because I didn't really need a wee, which totally threw both Virginia and myself. I talk about baby pressing on the bladder, which can cause some women to empty their bowels during labour or at least to feel like they want to do so. When this happened to me this time, because I felt as though I wanted to wee, I didn't realize that it was the baby moving down past my bladder and neither did Virginia. She later said that in all her time of midwifery, no one had ever said they wanted to wee at that stage before. So I'd got out of the pool and I was standing as I experienced this most intense surge, when I lost a lot of water and my bowels emptied! Yep, it was just a little bit but out it came as my body's natural expulsive reflex kicked into play.

It feels weird to write this here, but this stage totally took me by surprise. My body tensed, I felt another really strong surge and the baby's head began to emerge. I was stuck. I didn't quite know what to do at this point. Virginia told me to get back in the pool, but I didn't want to move. She calmly told me again to get back in before the next surge and Jerome helped me back into the water. I put my hand around the back of me and could feel the top of Aluna's head, although I didn't tell Virginia or Jerome.

I could feel all of her hair. I was still in a bit of shock. I remember feeling the tension in my body and I said to Jerome, 'It hurts', and I knew I had to surrender, so I finally made the decision to allow my body to sink into the water. As I did this, the next surge took over and Aluna's whole head came out. The hardest part was over.

I brought my left hand down between my legs and, as I surged again, guided Aluna out, and Virginia then brought her around to the front. Again, I think I was a bit shellshocked, as I didn't immediately pick her up out of the water (which is fine, as newborns don't breathe in until the cold air hits their face). After what probably wasn't longer than a couple of seconds, I lifted her to my chest and then came a rush of emotion that felt so different to my previous two births: the joy and relief that the oxytocin had done it. I had just had a quick, intense, powerful, beautiful birth and delivered my own baby, and there she was: full head of hair, beautiful brown eyes, just lying on my chest. It was amazing.

As always, the toughest part of my labour was delivering the placenta. Even though Aluna's birth had been super quick – Virginia had arrived at 4 a.m. and we had welcomed Aluna by 5 a.m. – it felt as though it had sapped every bit of energy and strength from me, and I made more of a fuss delivering the placenta than I had the baby! Go figure!

Lauren had arrived and was telling me to use my breathing and I remember thinking I couldn't breathe! But I did (of course) – the placenta came out and my job was finally done.

All that was left was for Aluna to start feeding and for me to get my lips around a lovely placenta smoothie. It wasn't until we had been out of the pool for at least

15–20 minutes that I even asked what sex 'it' was. Again, I was super shocked, as right throughout my pregnancy everyone had been telling me we were going to have a boy, so I thought we were having a boy. It was amazing to have such a lovely surprise. Coco and Caesar woke up just as Aluna started feeding, and they came downstairs, eyes wide, not quite believing what they were seeing. It was magical.

We didn't name Aluna until later that day. It was a toss-up between Luella and Aluna, but as Aluna was born at full Moon, and the name means 'Goddess of the Moon', it seemed very fitting.

I felt and feel so blessed to be able to share this totally magical experience with you. Birth is amazing and I look forward to helping many, many more women through The Calm Birth School Instructors across the UK and through the video course to enjoy calm and positive births.

References

1. Gas and air is also known as nitrous oxide and by the commercial name, Entonox.

2. Ayers, S., Eagle, A. and Waring, H. 'The effects of childbirth-related post-traumatic stress disorder on women and their relationships: A qualitative study', Psychology, Health & Medicine, 2006; 11(4): 389-98: http://dx.doi.org/10.1080/13548500600708409

3. Beck, C., Gable, R., Sakala, C. and Declercq, E. 'Posttraumatic stress disorder in new mothers: results from a two-stage U.S. national survey', Birth, 2011; 38(3): 216-7; doi: 10.1111/j.1523-536X.2011.00475.x. Epub 20 May 2011.

4. Also known as a 'membrane sweep' or stretch and sweep, this is a simple procedure in which a care provider will use their fingers to massage the neck of the cervix, in the hope of stimulating the uterus and kick-starting labour.

5. Propess is the commercial name for a slow-release pessary that contains prostaglandins and is inserted into the vagina to induce labour.

6. The practice of ingesting the placenta after it has been steamed, dehydrated, ground and placed into pills.

7. www.themindfulnessinitiative.org.uk/images/reports/Mindfulness-APPG-Report_Mindful-Nation-UK_Oct2015.pdf

8. www.huffingtonpost.com/mark-matousek/the-power-of-solitude_b_5276055.html

9. https://www.ncbi.nlm.nih.gov/pmc/articles/PMC3580050/

10. www.webmd.com/baby/features/childbirth-options-whats-best#3

11. Ibid.

12. Hodnett, E., Gates, S., Hofmeyr, G. and Sakala, C. 'Continuous support for women during childbirth', Cochrane Database of Systematic Reviews, 2013, Issue 7: DOI: 10.1002/14651858. CD003766.pub5

13. https://alaboroflove.org/sherpa-doula/

14. Alfirevic, Z., Devane, D. and Gyte, G. 'Continuous cardiotocography (CTG) as a form of electronic fetal monitoring (EFM) for fetal assessment during labour', Cochrane Database of Systematic Reviews, 2013, Issue 5: DOI: 10.1002/14651858.CD006066.pub2

15. https://www.ncbi.nlm.nih.gov/pmc/articles/PMC3163659/

16. www.kickscount.org.uk/

17. https://www.ncbi.nlm.nih.gov/pubmed/21280989

18. https://midwifethinking.com/2017/01/11/pre-labour-rupture-of-membranes-impatience-and-risk/

19. https://www.ncbi.nlm.nih.gov/pubmed/22592681

20. www.homebirth.org.uk/thirdstage.htm

21. https://www.rcm.org.uk/sites/default/files/Third%20 Stage%20of%20Labour.pdf

22. https://patient.info/doctor/retained-placenta

Additional Support
and Resources

Calm Birth School classes

If you would love to have support in person and find an instructor near you, please visit: www.thecalmbirthschool.com

The Calm Birth School video programme

While I offer a very limited number of face-to-face consultations per year, one way to experience the equivalent of a private class in the comfort and convenience of your own home is to invest in The Calm Birth School video programme. The benefits of this include:

- Being able to control and direct your learning by participating in the classes as many times as you want or need, right up until the day you go into labour.

- Learning in two different formats – reading and visual/ auditory via the video – which many people find suits their learning style.

- You might find it easier to engage your birth partner in the video programme.

All students of the video programme also get direct access to me through The Calm Birth School closed Facebook group, where I can personally coach and advise you through any questions, concerns or fears that may arise during your pregnancy journey.

All purchasers of *Calm Birth* can purchase The Calm Birth School course with a 15 per cent discount using the coupon code TCBS15BOOK.

Pregnancy and birth

AIMS (Association for Improvements in Maternity Services): Campaigns for better understanding of the normal birth process, and also provides support and information about maternity choices in the UK and Ireland.
www.aims.org.uk

Align Your Baby: Help for women whose babies are in non-optimal positions for birth.
https://alignyourbaby.com

BellyBelly: Pregnancy, birth and parenting website.
www.bellybelly.com.au

Birthrights: Promoting dignity and human rights in childbirth.
www.birthrights.org.uk

Birth Story Listeners: Peer support for mums in North Wales who have had a difficult or distressing experience of childbirth.
birthstorylisteners@gmail.com
www.facebook.com/groups/birthstorylisteners/

Birth Trauma Association: If you have not had the birth you wanted and need some impartial support to help you to work through your experience.
www.birthtraumaassociation.org.uk

Bump to 3: Not-for-profit organization offering support and courses to parents and parents-to-be.
Clair@Bumpto3.org.uk
www.Bumpto3.org.uk

La Leche League GB: Information and encouragement for breastfeeding mothers.
www.laleche.org.uk

Mama Academy: Empowering mums and midwives to help more babies arrive safely.
www.mamaacademy.org.uk

Tell Me A Good Birth Story: Database of 'birth buddies' offering free support for your unique situation.
www.tellmeagoodbirthstory.com

The Positive Birth Movement: Worldwide groups of women dedicated to sharing and supporting women in creating positive birth experiences.
www.positivebirthmovement.org

Further reading

Janet Balaskas, *New Active Birth* (Thorsons, 1991)

Dean Beaumont, *The Expectant Dad's Handbook* (Penguin Random House, 2013)

Sarah J. Buckley, *Gentle Birth, Gentle Mothering* (Penguin Random House, 2009)

Grantly Dick-Read, *Childbirth Without Fear* (Pinter and Martin, 2013)

Ina May Gaskin, *Ina May's Guide to Childbirth* (Ebury, 2008)

Kicki Hansard, *The Secrets of Birth* (self-published, 2015)

Mark Harris, *Men, Love & Birth* (Pinter and Martin, 2015)

Milli Hill, *The Positive Birth Book* (Pinter and Martin 2017)

Michel Odent, *Birth Reborn* (Pantheon, 1984)

The Calm Birth School Breathing Techniques at a Glance

The breathing techniques are all included here again for ease of reference.

∼ The Calm Birth School Breathing Technique ∼

TCBS breathing is a simple, powerful technique. The reason it is so effective is because it triggers the body's natural calming reflex, which occurs when the out-breath is nearly twice as long as the in-breath. During your labour, this technique will help you to maintain a deep state of calm and you'll use it in between surges. You'll also use it as you feel a surge coming in and once it has subsided.

Ideally, the breath is taken in and out through the nose as opposed to the mouth, as this gives you more control over the flow of air. However, please don't stress if you have a cold when you're birthing. Just breathe through your mouth – it will all be OK.

How to do it

Breathe in deeply to the count of four through the nose, imagining you're filling your lungs right to the bottom. As you breathe out, imagine sending the breath down, so it moves around your baby, down your legs into the tips of your toes and then into the floor. Sometimes, when you're first practicing this technique, it's useful to put your hands on either side of your waist, so you can feel the rise and fall as you breathe deeply. It really is as simple as that.

When to do it

Use this technique whenever you find yourself feeling stressed, whether it be at work, with your partner or getting on and off public transport – wherever and whenever. This will help you to relax quickly the more that you practise it. In addition, should you be one of the many of women who labour quickly, this technique will get you into the birthing zone quickly and easily once you have become accustomed to using it in your everyday life.

If you're the type of person who tends to take life in her stride with very little stress, this doesn't make you exempt from practising. It's just as important for you to carve out some practice time for your Calm Birth School breathing, too. Ensure you do one set in the morning for five minutes, five minutes at lunchtime and five minutes again in the evening.

～ Wave breathing ～

Use wave breathing when you're experiencing a wave or surge.

How to do it

The central idea of wave breathing is to keep both the inhalation and the exhalation even. Breathe in through the nose to the count of seven and out through the mouth for seven.

The role of this breath is to work with the upwards motion of the uterus as it rises, and then to send your breath down to your baby and your womb, while relaxing. Simple!

However, don't be fooled. For you to move instinctively into that space of deep breathing and relaxation when you experience a wave, you need to have practised it often beforehand so that you're used to it.

Please do not worry if you're unable to keep the breath even for a count of seven to start with. Work with whatever feels most comfortable for you. Perhaps you'll start off counting to four and once that feels good extend it to five. The main point is you start to feel comfortable slowing your breathing down and taking control of the flow. This will help you immeasurably during labour and birth.

When to do it

My suggestion is to practise this type of breathing for five full minutes every morning. If that means setting your alarm five minutes earlier – do it. It's such a good way to start your day and will leave you feeling great, as well as preparing you for your labour day, when you'll be using it during each surge you experience.

∼ Birth breathing ∼

Use this technique when you're experiencing a wave or surge and are fully dilated.

How to do it

The best way to aid the natural expulsive reflex is to work with the breath in a way similar to when you're wave breathing. The main difference is that you place all the emphasis on the exhalation. The out-breath needs to be very long and very deep. It can be helpful to use a visualization to accompany the out-breath – anything that reminds you of the importance of staying open, relaxed and moving downwards.

Some women use the words 'open, relaxed or release'; others will think about there being no resistance, or imagine a flower opening, or will picture something significant to them that helps to keep the idea of openness in their mind.

It can be really helpful to work with noise when you're getting to this stage. Although some will feel equally comfortable working with the breath alone, others will want to hum, shout, groan or even moo. It can get incredibly primal at this stage. This is nothing for either you or your birth partner to fear. No resistance is the main aim of the game. If you want to howl, just howl! Whatever you instinctively want to do is all-good – seriously. The only thing to be mindful of is to use the noise and the energy to send your power back down to your baby and your uterus, so they can finish the job.

When to do it

The best place to practise this technique is when you're having a poo. If you're at home, hum when you're on the loo so you start to feel more comfortable and familiar with directing your sound and energy down in that way.

This is great if you're suffering with constipation too. It won't shift everything immediately, but by applying patience, the humming and the breathing will see your natural expulsive reflex start to get things moving much more quickly and comfortably.

As mentioned above, use this technique when you're experiencing a wave or surge and are fully dilated.

Index

ABOUT THE AUTHOR

Philippa James Photography

Suzy Ashworth used hypnobirthing to bring her first child, Caesar, into the world. Following a second home water birth with her daughter, Coco, and after speaking to a plethora of women who shared their negative experiences of birth, Suzy developed The Calm Birth Method to help more women understand how they could empower themselves during pregnancy and birth to create more positive birth experiences.

Suzy believes that a woman's birth experience has the power to influence the way she feels about herself, her child and even who she is as an individual. Her Calm Birth Method aims to help women approach birth feeling confident, empowered and knowledgeable, and provides all the tools needed to ensure that a woman's birth experience is both calm and positive. As Suzy says, every mother is a birthing goddess and no one can fail at birth, regardless of the way her baby decides to enter the world.

Suzy holds diplomas in hypnotherapy and psychotherapy. She is a wife, sister, author, speaker, business mentor and mother of three beautiful children.

www.thecalmbirthschool.com

HAY HOUSE

Look within

Join the conversation about latest products,
events, exclusive offers and more.

f   Hay House UK

🐦   @HayHouseUK

📷   @hayhouseuk

❤   healyourlife.com

We'd love to hear from you!